T5-AFU-619

Lecture Notes
in Economics and
Mathematical Systems
Managing Editors: M. Beckmann, Providence, and H. P. Künzi, Zürich

Operations Research

29

Salah E. Elmaghraby
North Carolina State University
Raleigh, NC/USA

Some Network Models in Management Science

Reprint of the First Edition

Springer-Verlag
Berlin · Heidelberg · New York 1970

Original Series Title: Lectures Notes in Operations Research and Mathematical Systen

658.502
E48s
Cop. 2

Advisory Board
H. Albach · A. V. Balakrishnan · F. Ferschl
W. Krelle · N. Wirth

ISBN 3-540-04952-5 Springer-Verlag Berlin · Heidelberg · New York
ISBN 0-387-04952-5 Springer-Verlag New York · Heidelberg · Berlin

This work is subject to copyright. All rights are reserved, whether the whole or part of the material is concerne specifically those of translation, reprinting, re-use of illustrations, broadcasting, reproduction by photocopying machi or similar means, and storage in data banks.

Under § 54 of the German Copyright Law where copies are made for other than private use, a fee is payable to the publishe the amount of the fee to be determined by agreement with the publisher.

© by Springer-Verlag Berlin · Heidelberg 1970. Library of Congress Catalog Card Number 72-125499. Printed in German

TABLE OF CONTENTS

JAN 10 '78

HUNT LIBRARY
CARNEGIE-MELLON UNIVERSITY
PITTSBURGH, PENNSYLVANIA 15213

PROLOGUE

This manuscript gives an expository treatment of four network models: Shortest Path Problems, Maximum Flow Models, Signal Flow Graphs and Activity Networks. Emphasis throughout is on the basic concepts of each model and on its relevance and applicability to management science and operations research problems.

This manuscript can be considered as an attempt to answer the question that is often asked by practitioners and students in the fields of management science and operations research alike: what _is_ the theory of networks and what can it _do_ for me? A way to paraphrase this question is as follows: what operational systems have been modeled by networks, and how useful are such models?

To answer such a question in an exhaustive manner would necessitate a manuscript of considerable size. Therefore, we must be content with a rather brief, though hopefully illuminating, discussion of the subject.

By necessity, we shall not dwell on the graph-theoretic aspects of the models discussed. Furthermore, we limit our discussion to a few models which are most strongly related to management science. In this we are making an implicit distinction between 'graph theory' and 'network theory'. So let us make the distinction explicit: a _graph_ defines the purely _structural_ relationships between the nodes, while a _network_ bears also the _quantitative characteristics_ of the nodes and arcs. Consequently, in what follows we take a graph-theoretic problem to be a problem related to pure structure. On the other hand, a network problem is related to the quantitative characteristics defined on the graph.

For example, the following is a graph-theoretic question: In a connected graph G of n nodes and A arcs, what is the minimum number of nodes that must be removed to eliminate all the arcs in the graph (this is the well-known 'minimal cover' problem)? Notice that the answer to the problem is evidently dependent only on the structure of the graph. On the other hand, the following is a network problem: Given a graph G of n nodes and A arcs, and given that arc (i,j) is of 'length' d_{ij}, what is the shortest path between two given nodes s and t (this is the well-known 'shortest path problem')? Here, it is equally obvious that the answer depends not only on the structure of the graph but also on the very specific values (the lengths) defined on the arcs of the graph.

In limiting ourselves to network problems as related to management science we are not implying that there are no graph-theoretic problems in management science, or that such problems are not interesting and their models fascinating from a mathematical point of view. All that is being said, actually, is that space does not permit a reasonable development of both subjects, and that choice fell to network problems. The reader is urged to consult the excellent book by Busacker and Saaty[4] for a lucid exposition of graph theory and its applications.

The vehicle of exposition is by example, and this monograph treats four kinds of network models:

1. Shortest path problems
2. Maximum flow models
3. Signal flow graphs
4. Activity networks

In each case a function is defined on the arcs of the network, but the algebra for the manipulation of these quantitative measures is different from model to model. This is a key concept in network models: while the structure of several networks may be identical, the analysis of the functional relations defined on the network may be different for different models. Hence, the results of the analysis are different!

There are a few graph-theoretic concepts which are common to all four models, and it is advantageous to introduce such concepts here rather than repeat them in each subsequent model.

A graph is a collection of arcs and nodes. Each arc is a line segment that connects two nodes, and is usually designated by its terminal nodes unless more than one arc join the same two nodes and then a different designation must be devised to differentiate among them. An arc with its two end nodes is usually called a branch. Naturally, the nodes, i.e., the end points of the line segment, need not be distinct: a self-loop in which the two end points are the same is clearly a legitimate branch. The nodes are sometimes referred to as vertices or junction points. A linear graph is a collection of branches that have no points in common other than nodes. A graph in which it is possible to reach any node from any other node along branches in the graph is said to be connected; otherwise it is unconnected. A chain between nodes i and j is a connected sequence of branches that lead from i to j such that each node is encountered only once. An arc may have an orientation, and then it is a directed arc or an arrow. A directed chain has all branch orientations leading from i to j. On the other hand, a directed branch sequence from i to j in which some of the branches are traversed opposite to their orientation is called a path. Thus a path may contain forward as well as reverse branches; a chain contains only forward branches.[15] Whenever the context of the problem does not permit the existence of paths with reverse branches, we shall use path synonymously with a chain. A chain whose terminal nodes coincide is called a loop or cycle. Notice the difference between a loop and a self-loop mentioned previously. A spanning tree is a connected subgraph† of all nodes which contains no loops. Obviously, every connected graph must have at least one spanning tree. And conversely, a spanning tree could not exist unless the original graph were connected. It is equally evident that if the graph G contains n nodes, the tree must contain exactly n - 1 arcs. A graph G may have several spanning trees. For any spanning tree, the arcs of G that are not in the tree are called chords. The order of a node is the number of arcs incident on it, i.e., the number of arcs for which the node constitutes one of their terminal points.

---

† A subgraph $H = (N',A')$ of a graph $G = (N,A)$ is a graph whose $N' \subseteq N$ and $A' \subseteq A$.

CHAPTER 1

THE SHORTEST PATH PROBLEMS

Contents

§1.1 INTRODUCTION

The first image that comes to mind when the word 'network' is mentioned is a traffic network, whether it be road or air traffic. Most of us are familiar with such networks since one rarely travels from one location to another without consulting a 'map', which is, in our terminology, a 'network'.

There are several interesting, and rather important, questions that can be raised concerning these networks, the most natural of which is: what is the shortest path from one location to another? This has come to be known as the 'Shortest Path Problem' which, together with its extensions and complications, constitute an interesting field of study in mathematical programming and combinatorial analysis. The subject is too wide for a brief coverage. Hence, this section can only aspire to demonstrate the kind of fundamental questions asked and the rather ingenious methods suggested for the resolution of a few of them, hoping that this will indicate the general sweep of the theory as well as arouse the reader's curiosity to find out the rest of the story. (In this connection, the reader is referred to the excellent survey paper by Dreyfus, ref. [10]).

We shall be interested in the problem of finding the shortest path between an origin, generically designated by s, and a terminal, generically designated by t, in a network of N nodes and A arcs. (It is worthy of note that while the *shortest* path problem has been extensively treated in the literature, and boasts several algorithms of varying computing efficiencies, the problem of finding the *longest* path which contains no loops is still largely unanswered except in some very special cases, because most networks do contain cycles of positive length.)

It is important to emphasize that while each arc (i,j) in the network bears a real finite number, d_{ij}, (usually $d_{ij} > 0$, but may be negative), which we shall refer to as the 'distance' between nodes i and j, the real physical meaning of this number may be anything, depending on the application. For example, it may represent time (in hours), or cost (in dollars), or reliability (expressed as probability of non-failure), or some other measure of interest. Our use of the terminology appropriate for physical distances is merely for concreteness of discussion. In fact, at the end of this section, we shall discuss the applicability of the algorithms developed here to other problems in which d_{ij} is certainly not physical distance.

At the outset, we eliminate a pathological case in which it is possible to drive the length of the shortest path between s and t to $-\infty$. Clearly, such an instance, while possibly of some theoretical interest, is of no practical value. It occurs if there is a cycle C of length $L(C) < 0$, and if there is a path Π from s to t *via* a node w incident with C. For then we could traverse Π from s to w, then go around C a sufficient number of times, and finally return to Π from w to t. In this manner we could drive the length of the path from s to t as small as we desire. For this reason, we shall assume that

$$L(C) \geq 0$$

for every cycle C in the network.

An interesting case in which the length of a cycle is negative has been reported in the exchange of foreign currencies. Due to a difference between the 'official' and the 'black market' rates of exchange, it is theoretically possible to amass a fortune by starting with 'hard currency' and ending with the same currency. We say "theoretically possible" because bankers are usually aware of these discrepancies and would block any exchanges beyond a certain 'reasonable' limit.

Generally speaking, one can classify problems of shortest path into four main categories:

Problem I. Find the shortest path between two specified nodes s and t.

Problem II. Find the m-shortest paths between two specified nodes s and t.

Problem III. Find the shortest paths from an origin s to all other nodes of the network (or alternatively, the shortest paths from all nodes to a terminal t).

Problem IV. Find the shortest paths between all pairs of nodes.

Problem II is a generalization of I and requires some explanation. Suppose we enumerate all the paths from s to t and compute their lengths. Suppose, further, that we rank the paths in ascending magnitude of their lengths, such that we have $L(\Pi_1) \leq L(\Pi_2) \leq \ldots \leq L(\Pi_m)$. Path Π_1 is called the shortest path between s and t; path Π_2 the second shortest; and so on. Our task, in Problem II, is to determine the first m-shortest loopless paths from s to t. This is not a trivial problem since we wish to avoid complete enumeration of paths from s to t, as was suggested above.

As for Problem III, it turns out that almost all algorithms that solve Problem I also solve Problem III. It seems inherent in combinatorial approaches for the solution of the shortest path between s and t to simultaneously evaluate the shortest paths between s and all other nodes of the network. Hence, no more will be said about this problem.

Problem IV, that of determining the shortest paths between all pairs of nodes is evidently a generalization of Problem I, when one lets the pair (s,t) range over all such possible pairs. This problem is intriguing from a computational point of view; for it is clear that repeated application of the algorithm for Problem I would eventually provide the answer to Problem IV. But consider the computing burden: if the network is directed (or even mixed) one needs to solve N(N-1) shortest path problems. Admittedly, this figure is reduced by one half in in the case of undirected networks (due to the symmetry of the distance matrix). Still, this is not an insignificant amount of calculation, and a pertinent question is: can we do better? We shall discuss below a method due to Floyd[14] which solves the problem in <u>one</u> sweep and requires only N(N-1)(N-2) <u>elementary</u> operations!

This section terminates with a discussion of 'Related Topics'. These are topics in management science, operations research and industrial engineering which are related to the Shortest Path problem in all its forms. A fundamental understanding of the Shortest Path problem adds greater insight into, and facility with, these problems and their solution.

§1.2 THE SHORTEST PATH BETWEEN TWO GIVEN NODES s AND t

Algorithms for obtaining the shortest path between the origin s and the terminal t (as well as between s and all other nodes reachable from s) are numerous; see Bellman and Kalaba[2], Dantzig[7], Ford and Fulkerson[15], Minty[22]† and Moore[23], among others. We shall present an algorithm due to Dijkstra[9] which has the advantages that it: (i) requires $3n^3$ elementary operations, where each operation is either an addition or a comparison, and hence is more efficient than the other algorithms; (ii) can be applied in the case of non-symmetric distance matrices, with positive and negative arc lengths, hence it is quite general; and (iii) does not require storing all the data for all arcs simultaneously in the computer, irrespective of the size of the network, but only those for the arcs in sets I and II (described below), and this number is always less than n; hence it is quite economical in its demands on computer memory.

The algorithm follows. It capitalizes on the fact that if j is a node on the minimal path from s to t, knowledge of the latter implies knowledge of the minimal path from s to j. In the algorithm, the minimal paths from s to other nodes are constructed in order of increasing length until t is reached.

In the course of the solution the nodes are subdivided into three sets:

A. the nodes for which the path of minimum length from s is known; nodes will be added to this set in order of increasing minimum path length from node s.

B. the nodes from which the next node to be added to set A will be selected; this set comprises all those nodes that are connected to at least one node of set A but do not yet belong to A themselves;

C. the remaining nodes.

The arcs are also subdivided into three sets:

I. the arcs occuring in the minimal paths from node s to the nodes in set A;

---

† As far as we can discern it seems that both G. J. Minty and G. B. Dantzig have formulated independently the same algorithm. Minty's version appeared in the review paper of Pollack and Wiebenson [29], in which they credited a private communication with Minty. Dantzig's version appeared in a paper [7], and later in his book, ref. [8].

II. the arcs from which the next arc to be placed in set I will be selected; one and only one arc of this set will lead to each node in set B;

III. the remaining arcs (rejected or not yet considered).

To start with, all nodes are in set C and all arcs are in set III. We now transfer node s to set A and from then onwards repeatedly perform the following steps.

Step 1. Consider all arcs a connecting the node just transferred to set A with nodes j in sets B or C. If node j belongs to set B, we investigate whether the use of arc a gives rise to a shorter path from s to j than the known path that uses the corresponding arc in set II. If this is not so, arc a is rejected; if, however, use of arc a results in a shorter connection between s and j than hitherto obtained, it replaces the corresponding arc in set II and the latter is rejected. If the node j belongs to set C, it is added to set B and arc a is added to set II.

Step 2. Every node in set B can be connected to node s in only one way if we restrict ourselves to arcs from set I and one from set II. In this sense each node in set B has a distance from node s: the node with minimum distance from s is transferred from set B to set A, and the corresponding arc is transferred from set II to set I. We then return to step 1 and repeat the process until node t is transferred to set A. Then the solution has been found.

As an example, consider the network of Fig. 1-1 and the associated worksheet of Table 1-1 which is self-explanatory. As can be seen, the result was obtained in five iterations and required only 10 comparisons.

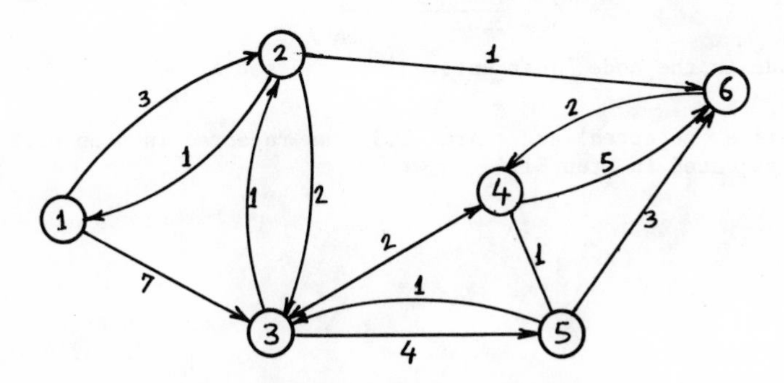

Figure 1-1

	A†	B	C	I	II*	III*
Start	(1)		1,2,3, 4,5,6			(1,2) (2,1) (3,2) (1,3) (2,3) (3,4) (4,3) (2,6) (3,5) (4,5) (5,3) (6,4) (4,6) (5,6)
S1.1	(1)	2,3	4,5,6		(1,2), (1,3)	
S2.1	1,(2)		4,5,6	(1,2)		
S1.2	1,(2)	3,6	4,5	(1,2)	(2,3), (2,6)	
S2.2	1,2,(6)	3	4,5	(1,2), (2,6)	(2,3)	
S1.3	1,2,(6)	3,4	5	(1,2), (2,6)	(2,3), (6,4)	
S2.3	1,2,(3), 6	4	5	(1,2), (2,3), (2,6)	(6,4)	
S1.4	1,2,(3), 6	4	5	(1,2), (2,3), (2,6)	(6,4), (3,5)	
S2.4	1,2,3, (4),6		5	(1,2), (2,3), (2,6), (6,4)	(3,5)	
S1.5	1,2,3, (4),6	5		(1,2), (2,3), (2,6), (6,4)	(3,5), (4,5)	
S2.5	1,2,3, 4,(5),6			(1,2), (2,3), (2,6), (6,4), (4,5)	(3,5)	

TABLE 1-1

† The circled node is the node 'just entered' in the set A

* A crossed arc is a 'rejected' arc. Arc (1,3) was rejected in step S1.2 and arc (3,4) was rejected in step S1.4.

§1.3 THE k-SHORTEST PATHS FROM s TO t

There are several good reasons for being interested in the k-shortest paths between two nodes, rather than just *the* shortest. For one thing, infeasibility of use of the shortest path may necessitate the adoption of the second shortest path; which, in case of *its* unavailability may necessitate the adoption of the third shortest path, etc. It may be advisable to have such alternatives evaluated and at hand to be used when needed. And, for another, the evaluation of the first k-shortest paths is some form of sensitivity analysis -- a test of how much a deviation from the optimal would cost. This is oftentimes of interest for purposes of network evaluation or activity analysis, as is more fully discussed in treatises on Activity Networks, see ref. [11].

There have been several attempts to determine, in an efficient manner, the k-shortest paths in a network. Some of these attempts proved later to be either incorrect or incomplete; see the discussion by Pollack[27],[28]. On the other hand, Hoffman and Pavley[18] have described a procedure which, unfortunately, does not show how to avoid paths with loops. We shall base our discussion on the paper by Clarke, Krikorian and Rausen[6]. Their procedure seems to be fairly efficient and complete. They report that a computer program for the UNIVAC 1107, a large high speed scientific machine, was expected to handle a 1023-node network with $k \leq 29$ in five minutes or less.

The procedure described here is capable of determining the k-shortest paths in any network, whether directed or undirected, with or without cycles, in which d_{ij} need not equal d_{ji}. We impose only one restriction: if the origin *s* is given, then any node of the network is reachable by a path from *s*. This condition is always satisfied in undirected networks, or directed networks in which each node is connected to some other node by two arcs in opposite directions. The special case of directed acyclic networks (of the PERT-CPM variety) will be treated separately below. There, we shall find that a much more efficient algorithm can be utilized.

Preliminaries.

First, we need a few definitions and basic relationships. A path is loopless if it does not include closed loops, i.e., if its nodes are all distinct. A path from i to j is called minimal if it is not longer than any other path from i to j. By necessity, a minimal path must be loopless, for if it has a loop removal of the loop shortens the path (we eliminate from our discussion loops of zero length since they possess no practical significance).

Once the origin s has been chosen, the base of the graph relative to s is defined to be the set of all arcs of minimal paths from s to all other nodes of the network. An arc is called basic if it belongs to the base. Notice that only in the case of a unique minimal path from s to every other node will the base be a tree, see Fig. 1-2.

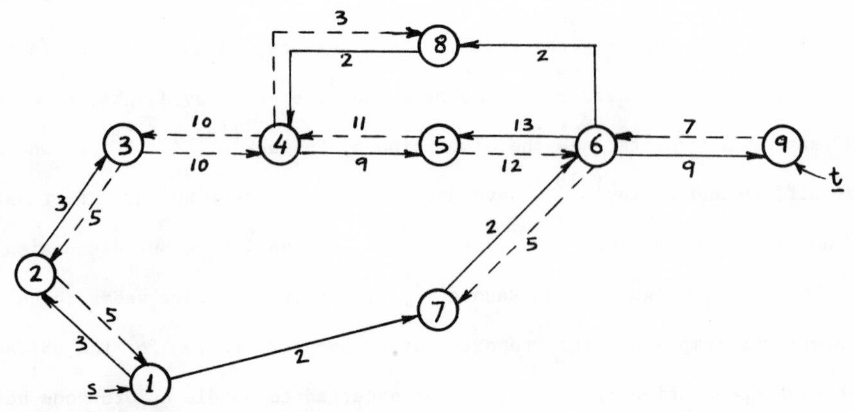

Figure 1-2. Heavy lines are basic arcs; dashed lines nonbasic.
Path 1,2,3,4,5,6,9 is loopless of degree 2 and length 46.
Path 1,7,6,9 is minimal, hence loopless, of length 13.
Path 1,7,6,8,4,5,6,9 contains a loop (consequently not minimal).

In the process of determining the base (by the Dijkstra Method discussed in §1.2, for example), each node j is assigned a label δ_j equal to the length of the minimal path from s to j. Clearly, a path Π from s to j is minimal if, and only if, $\delta_j = L(\Pi)$. That is to say, a path Π from s to j is minimal if, and only if, the label of j is equal to the length of the path.

If we define a section p of a path Π as any subpath contained in Π, then it immediately follows that the length of every section of a minimal path is minimal. For if this were not the case, suppose p joins nodes i and j, and let $\bar{p}$ be a minimal path between i and j. Then substituting $\bar{p}$ instead of p would yield a shorter path between the original terminal nodes of Π; see Fig. 1-3. This, however, contradicts the minimality of Π.

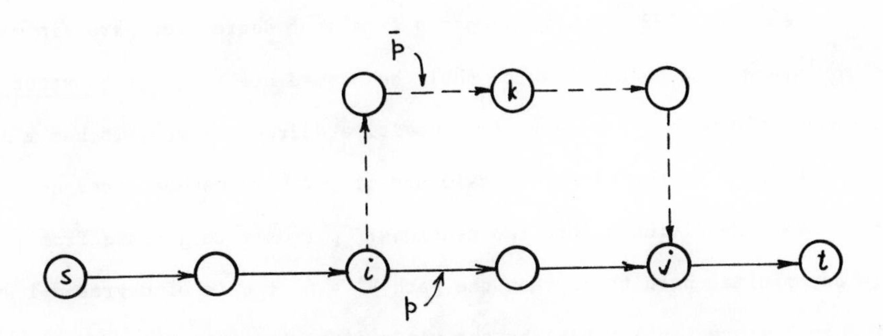

Figure 1-3. Heavy lines denote path Π. If $L(\bar{p}) < L(p)$, then the path s,i,k,j,t is a path between s and t which is shorter than Π.

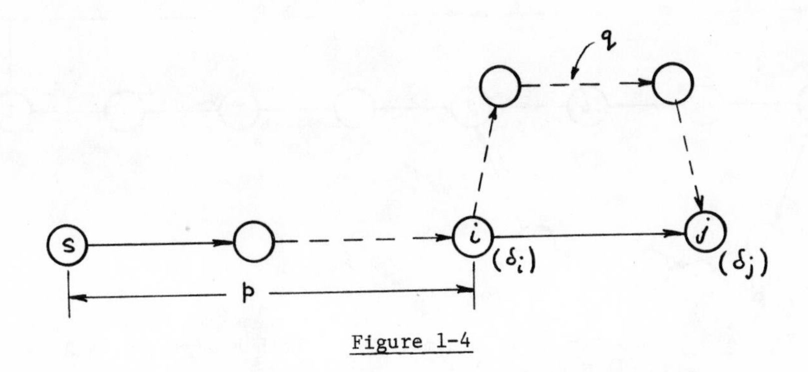

Figure 1-4

Returning to the concept of basic arcs, we immediately see that an arc (i,j) is basic if and only if $\delta_j = \delta_i + d_{ij}$. To see this, notice that if (i,j) is basic then (i,j) must lie on the minimal path from s to j, which can be decomposed into the subpath p from $s \rightarrow i$ and the arc (i,j). Since p is a section of a minimal path, p itself is minimal, as we have just shown. If $\delta_j \neq \delta_i + d_{ij}$, then there must be another subpath q which is minimal between i and j; see Fig. 1-4, and in this case arc (i,j) is not basic, which contradicts the assumption.

This clearly shows that the length of a minimal path is the sum of its sections, which themselves are minimal. Moreover, a minimal path p between i and j is composed exclusively of basic arcs, and conversely, any path composed exclusively

HUNT LIBRARY
CARNEGIE-MELLON UNIVERSITY
PITTSBURGH, PENNSYLVANIA 15213

of basic arcs defines a minimal path.

Any path Π from s to k is a sequence of arcs some of which may be basic and some may not. By the <u>degree of a path</u> shall be meant the number of nonbasic arcs in the path; see Fig. 1-2. Thus, all paths from s of degree zero are minimal paths.

Two interesting concepts which shall be needed are those of <u>shortcut</u> and <u>detour</u>. Consider a path Π from s to t of positive degree m (i.e., it has m nonbasic arcs) and let (i,j) be the first nonbasic arc of Π (there may be other nonbasic arcs if $m \geq 2$). We divide Π into two sections: p from s to j and q from j to t. If p' is any minimal path to j, then the path Π' = p' + q is of degree m-1 and is shorter than Π since L(p') < L(p) by the minimality of p' and the nonbasic character of (i,j). A path Π' constructed in this manner shall be called a shortcut of Π. Conversely, Π is a detour of Π'. Clearly, L(Π) > L(Π'); see Fig. 1-5.

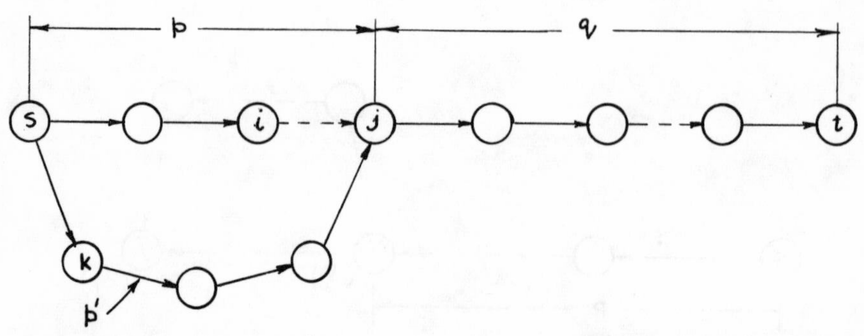

<u>Figure 1-5</u>. Subpath Π' is minimal to j. Path Π = p+q; path Π' = p'+q; m(Π) = 2, m(Π') = 1.

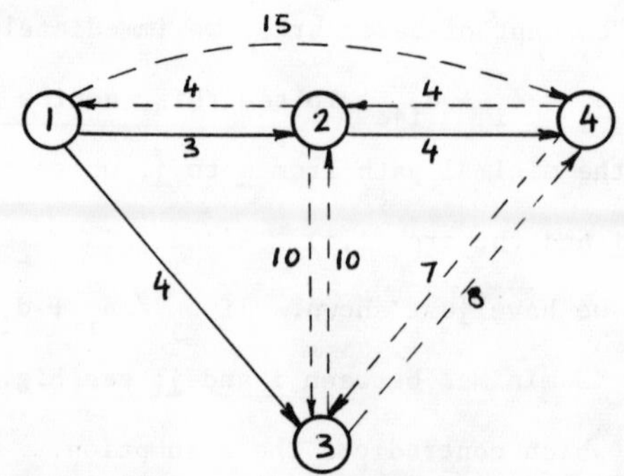

<u>Figure 1-6</u>. Path Π = 1,2,3,4 is of degree m=2 (arcs (2,3) and (3,4) are nonbasic). It has no detour of degree 3. But it has a short cut of degree 1 (path 1,3,4), which is not minimal.

Every path from s of degree $m > 0$ has a shortcut, because there is a minimal path from s to every other node in the network. But every path need not have a detour, i.e., a path of degree $m \geq 0$ need not possess a detour of order m+1; see Fig. 1-6.

If Π is loopless, a shortcut of Π may not be loopless. We saw an example of this in Fig. 1-2, in which Π = 1,2,3,4,5,6,9; p= (1,2,3,4) q= (4,5,6,9), Π = p + q; the shortcut is p'= 1,7,6,8,4 and the resulting path Π'= p' + q is not loopless. However, the loops that may occur in Π' are of a simple type possessing a unique property, which will be described presently since it is used in the algorithm.

Relative to any path Π from s we wish to define two sections constructed from Π in a special manner: the root, r(Π), and the spur, s(Π), such that Π = r(Π) + s(Π), and the first arc of s(Π) is the first nonbasic arc of Π; see Fig. 1-7.

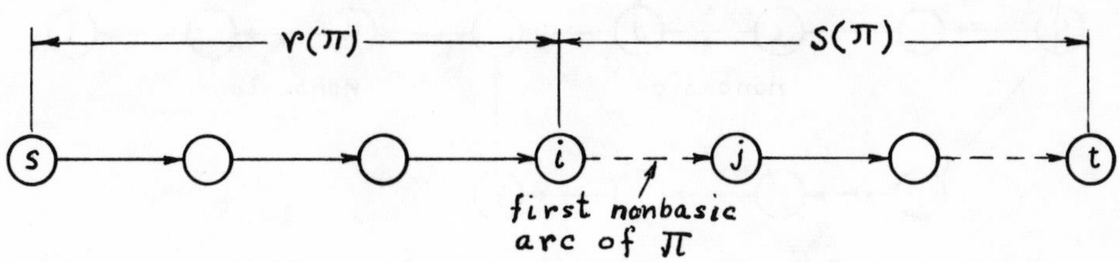

Figure 1-7. Definitions of root and spur.

Naturally, r(Π) is minimal since it is composed of basic arcs. If Π itself is minimal it coincides with r(Π), and s(Π) is empty. If the first arc of Π is non-basic, r(Π) is empty and s(Π) = Π.

Since we are interested in loopless paths and, as we shall see presently, the algorithm for constructing the k-shortest paths is iterative in nature starting from the terminal node t, let us define a path to be admissible if its spur is loopless. We shall be investigating only admissible paths.

It follows that if Π' is a shortcut of Π, Π' is admissible if Π was admissible. This can be readily seen with reference to Fig. 1-8. If Π is admissible then, by definition, s(Π) = (i,j) + q is loopless. If p' is the minimal path to j, then Π' = p' + q must also be admissible since s(Π') is given by the path k $\rightarrow$ t

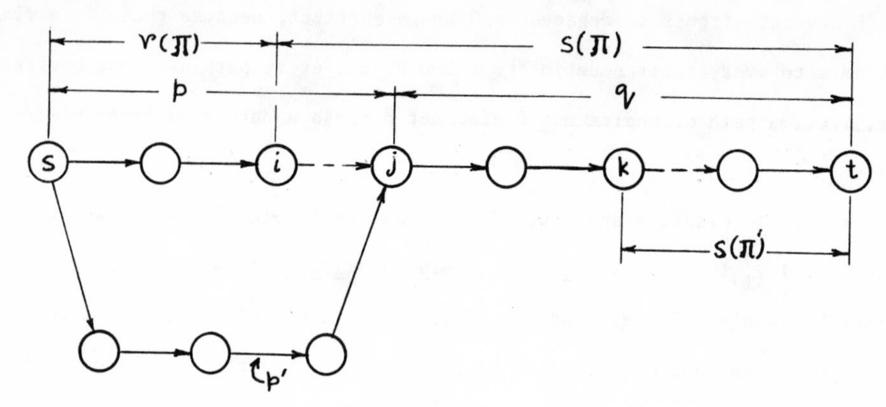

Figure 1-8.

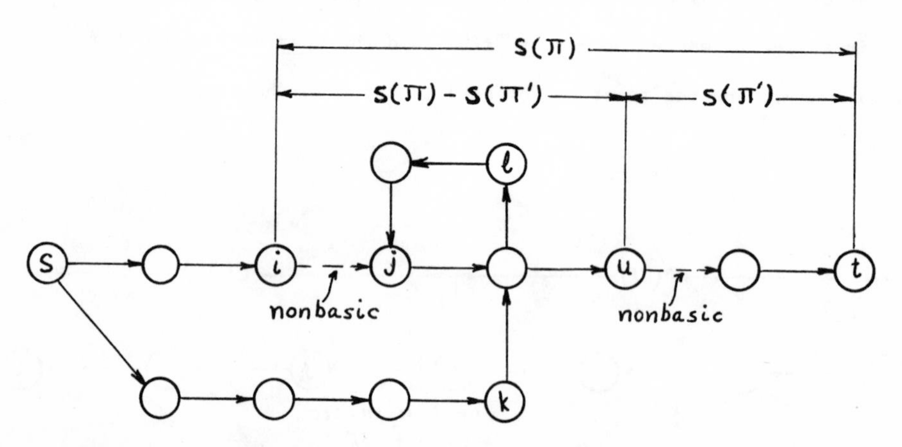

Figure 1-9. Π = $\underline{s}\to\underline{i}\to\underline{j}\to\underline{t}$; $r(\Pi)$ = $\underline{s}\to\underline{i}$; $s(\Pi)$ = $\underline{i}\to\underline{t}$.
Π' = $\underline{s}\to\underline{k}\to\underline{\ell}\to\underline{j}\to\underline{t}$; $r(\Pi')$ = $\underline{s}\to\underline{u}$; $s(\Pi')$ = $\underline{u}\to\underline{t}$
Π' contains a loop between $\underline{k}$ and $\underline{u}$, though it is admissible.

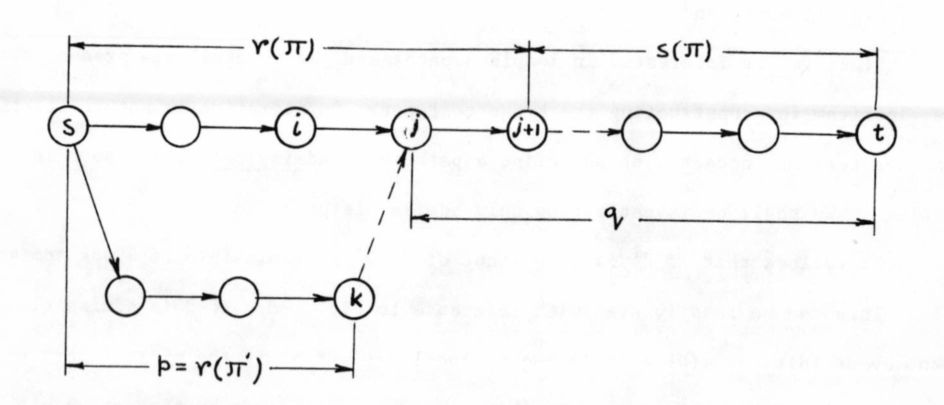

Figure 1-10

which is a section of q, and the latter was assumed loopless.

However, while Π' is admissible, i.e., its spur is loopless, Π' itself may not be loopless, as remarked above. Obviously, this can happen only if the loop occurs prior to $s(\Pi')$, i.e., in $r(\Pi')$. But, by assumption, $s(\Pi)$ is loopless. Therefore, the node that is repeated in Π' must be in both $r(\Pi')$ and $s(\Pi)$ - (i,j), and it is unique. This explains our statement above that if a shortcut path is not loopless, the loop is of a simple type. The situation is illustrated in Fig.1-9 (also see Fig. 1-2).

We are now in a position to use the above notions of loopless paths, admissible paths, and detours in constructing the algorithm for determining the k-shortest paths.

The Algorithm

Let A_o be the set of minimal paths from s to t and, for any positive integer m, let A_m be the set of all admissible paths of degree m from s to t. Then, for $m \geq 1$, A_m consists of all admissible detours of paths in A_{m-1} because, by the above discussion, every path in A_m has an admissible shortcut which must be a path in A_{m-1}. And conversely, any admissible detour of a path in A_{m-1} is of degree m and therefore in A_m.

The procedure divides itself naturally into three stages, the first two of which are simple and straightforward, the third stage being the most laborious. They are:

Stage 1. The determination of A_o: This is accomplished, as mentioned before, by the Dijkstra Method or any other algorithm referenced in §1.1. At the end of this stage we have the set of minimal paths from s to t.

Stage 2. The construction of the set S_1: The set S_1 is a set of n paths from $s \rightarrow t$ which are all loopless, and serves as the first approximation to the desired result (the set of k-shortest paths), S_{min}.

The manner in which S_1 is constructed is somewhat arbitrary, though systematic. Starting with the set A_o, if the number of paths $\geq$ k, choose any k of them and these are the desired k-shortest paths. S_{min} is at hand, and we are done. If not, construct A_1, the set of admissible detours of paths in A_o, and separate the loopless paths in A_1 from the detours with loops. If the total number of loopless paths is $\geq$ k, choose the shortest k among them and go to Stage 3. Otherwise if the total number of paths in A_o and A_1 is still less than k, construct A_2 in a similar fashion to A_1, etc. Eventually, sets $A_o, A_1, \ldots, A_k$ are constructed such that the total number of paths is $\geq$ k. Choose the shortest k paths among them; let this set be called S_1. S_1 is the first approximation to the desired set of k-shortest paths, S_{min}. The manner in which S_1 is refined is detailed in Stage 3 of the procedure. For the moment, let M be the maximum length of the paths in S_1. Keep all admissible, but not loopless, paths previously evaluated whose length is $< M$ as elements of another list of paths, call it T, since some of these paths, as well as the paths in S of length $< M$, may have loopless detours of length $< M$, which would be substituted for longer paths in S_1, resulting in an improved version of S_1.

Stage 3. The refinement of S_1 to yield S_{min}: We recall the manner in which an admissible detour is constructed. Let Π be an admissible path. Then Π' is a detour if it is of the form: $\Pi' = p + (k,j) + q$, where (i,j) is an arc of $r(\Pi)$, (k,j) is a nonbasic arc of the network, p is a minimal path to k, and q is the subpath identical with the section of Π from j to t; see Fig. 1-10. Furthermore, Π' is admissible if two conditions are satisfied: (i) the section of Π from j to t is loopless and, (ii) k is not one of the nodes j+1 to t. By construction, the section p of Π' is equal to $r(\Pi')$, the root of the detour path Π'. Moreover, since roots of paths are minimal paths, two paths which differ only in their roots are of equal length (because the two roots are equal in length, both being minimal paths to the same node, and the two paths share the same spur thereafter). And if one path is admissible, the other must be admissible also (because it has the same spur). Because of these two shared properties, we group such paths together and call them equivalent. In other words, equivalent paths differ only in their roots, have the same length, have the same degree (i.e., number of nonbasic arcs), and are either all admissible or all inadmissible.

Consequently, in constructing admissible detours, it is sufficient to provide a method for constructing and testing all admissible detours of an equivalent class of admissible paths from s to t. This necessitates the grouping of admissible paths into equivalent classes from the beginning.

The procedure now boils down to the following. Consider any class of equivalent paths. Let $w_o = (j, j+1, \ldots, t)$ be the common spur of all paths in the class; w_o may be empty if the class is of degree zero. Store w_o (together with its length) as the first element of a table W. Consider all arcs (k,j) incident on node j. Reject arc (k,j) for any of the following five reasons:

(R1) $\delta_k = \delta_j + d_{kj}$, i.e., node j lies on the minimal path to k.

(R2) $k \varepsilon w_i$; i.e., node k lies on the section j to t, because then the new spur is inadmissible (has a loop).

(R3) (k,j) nonbasic but path $p(k)+(k,j)+w_i$ is in S or T; where p(k) is the minimal path to k; i.e., the new path via k is already in the set S or T.

(R4) $\delta_k + d_{kj} + L(w_i) \geq M$; i.e., the length of the new path is $\geq$ M.

(R5) p(k) has a node other than k in common with w_i, because then a loop will be formed.

If k is not rejected for any of the above reasons then either

(i) (k,j) is nonbasic, and then a family of equivalent detours is obtained by taking paths of the form $p(k) + (k,j) + w_o$ for all minimal paths p(k) from s to k. By construction, all these paths are of length less than M. They are added to S if they possess no loops, eliminating from S an equivalent number of longest paths, and recomputing M. If any path possesses a loop, it is added to T and all paths of length larger than the recomputed M are eliminated.

(ii) (k,j) is a basic link. If w_i was the last path entered in the table W, let $w_{i+1} = (k,j) + w_i$, and add it to W (together with its length).

After all the arcs incident on j have been tested, repeat the procedure on the initial node of w_1, then that of w_2, and so on. The testing of this class of equivalent paths stops when, for some nonnegative integer v, after testing all arcs which are incident on the initial node of w_v one finds no next entry w_{v+1} in the table W.

The process is then repeated with another class of equivalent paths in S, and so on until all classes have been exhausted.

Numerical Example

Consider the undirected network shown in Fig. 1-11, in which $\underline{s}=\underline{1}$ and $\underline{t}=43$, and we are interested in the k=10 shortest paths.

Stage 1. Application of the Dijkstra Method results in the basis shown in heavy lines in Fig. 1-11, and the labeling δ_j shown next to each node. Notice that the basis is not a tree because of the existence of the loop (26, 27, 43).

Stage 2. To construct the set S_1 we chose arbitrary detours yielding loopless path The set S_1 and the lengths of paths are shown in Table 1-2. The circled entries denote the start of the spur of each path. Notice that paths 1 and 2 belong to th same class, (i.e., they are equivalent), since they are minimal paths to $\underline{t}$ and therefore possess the same (empty) spur. All other paths differ in their spurs and consequently belong to different classes. From Table 1-2, M = 148.

Stage 3. The iterations are simplified if the format of Table 1-3 is followed. I Table 1-3 we carried out sample calculations until a new path was generated, which replaced the longest path in S_1 and resulted in S_2. The complete changes in S_1, together with other pertinent information, are summarized in Table 1-4. The reader can easily analyze the remaining classes of paths in Table 1-4 and verify that the 10 paths shown are indeed S_{min}, the 10-shortest paths from node $\underline{1}$ to node $\underline{43}$.

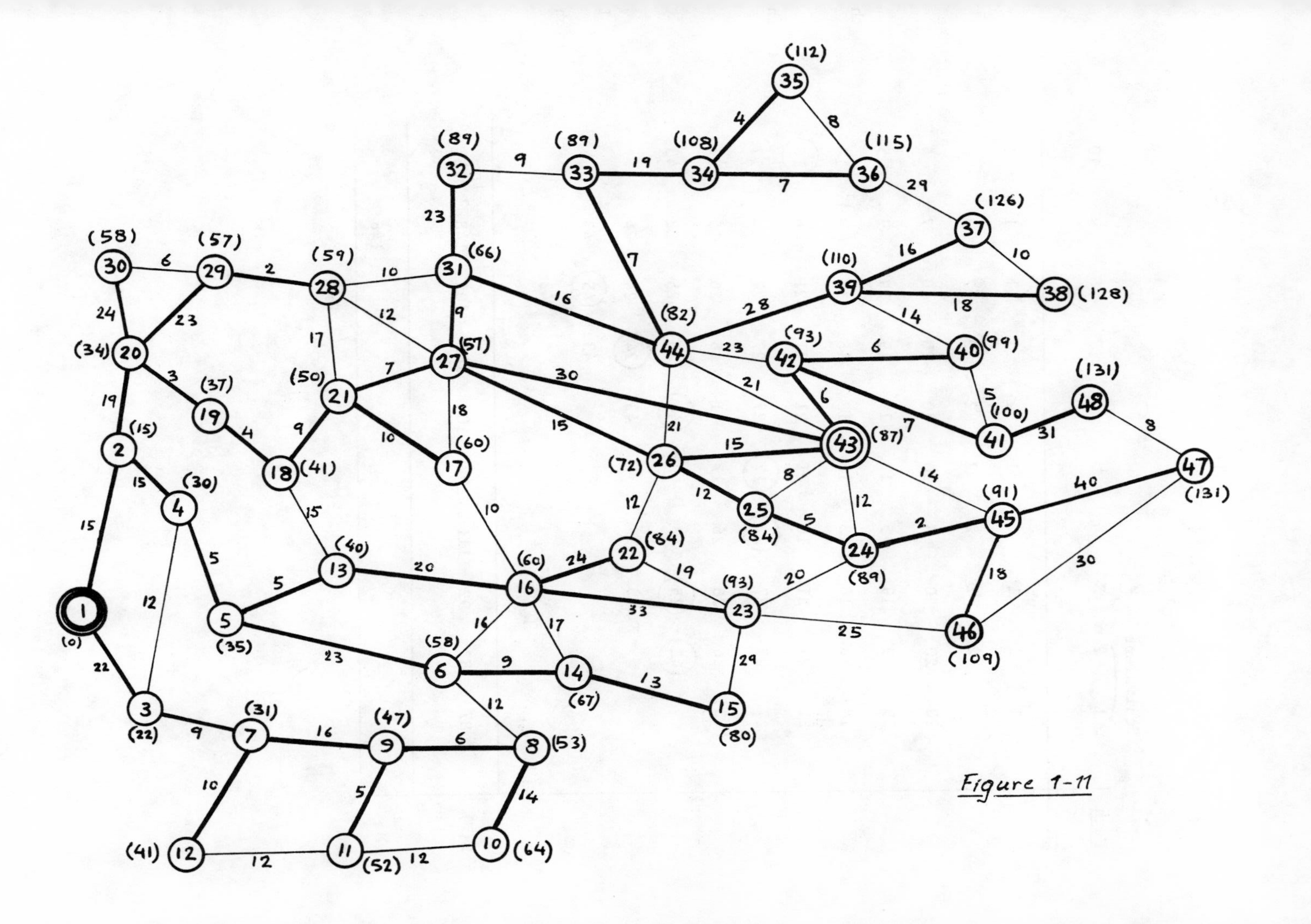

Figure 1-11

TABLE 1-2

	Equivalent									
Path	1	2	3	4	5	6	7	8	9	10
	1	1	1	1	1	1	1	1	1	1
	2	2	2	2	2	2	2	2	3	2
	20	20	20	4	20	20	20	20	7	4
	19	19	19	5	19	19	19	19	9	5
	18	18	18	13	18	18	18	18	8	13
	21	21	21	16	21	21	21	21	6	16
	27	27	27	22	17	27	27	27	14	17
	43	26	26	26	27	31	26	26	15	27
		43	25	43	43	44	25	25	23	43
			43			43	24	24	24	
							43	45	43	
								43		
L(Π)	87	87	92	111	108	103	101	105	148	117
M									148	

TABLE 1-3

	First Node of w_i: $\underline{j}$	Connecting Node $\underline{k}$	Reason for Rejecting (k,j)	Accept (k,j)			
				(k,j) Basic w_{i+1}	$L(w_{i+1})$	(k,j) non-basic	$L(\Pi)$
o	43	24	R3				
		25	R3				
		26		$(26,43)\equiv w_1$	15		
		27		$(27,43)\equiv w_2$	30		
		42	R1				
		44	R3				
		45	R3				
1	26	22	R3				
		25	R1				
		27		$(27,26,43)\equiv w_3$	30		
		43	R1				
		44				$(1,2,20,19,18,21,27,31,44,26,43)\equiv\Pi_{11}$	118

11 replaces Π_9 in Table 1. The new M = 118. Continue.

TABLE 1-4 Summary of Changes $S_1 \to S_{min}$

	Starting S_1										Changes					
Path	1	2	3	4	5	6	7	8	9	10	11	12	13	14	15	16
	1	1	1	1	1	1	1	1	1	1	1	1	1	1	1	1
	2	2	2	2	2	2	2	2	3	2	2	2	2	2	2	2
	20	20	20	4	20	20	20	20	7	4	20	20	20	20	4	4
	19	19	19	5	19	19	19	19	9	5	19	29	19	29	5	5
	18	18	18	13	18	18	18	18	**8**	13	18	**28**	18	**28**	**13**	**13**
	21	21	21	16	21	21	21	21	6	**16**	21	27	21	27	18	18
	27	27	27	**22**	**17**	27	27	27	14	17	27	43	**17**	26	21	21
	43	26	26	26	27	31	26	26	15	27	31		27	43	27	27
		43	**25**	43	43	**44**	25	25	23	43	**44**		26		43	26
			43			43	**24**	24	24		26		43			43
							43	**45**	43		43					
								43								
L(Π)	87	87	92	111	108	103	101	105	148	117	118	101	108	101	101	101
M									148		118	117	108	108	108	105
Equiv. to	2	1	-	-	-	-	-	-	-	-	-	-	-	-	-	-
Replacing path No.											9	11	10	4	5	13
Determined from w_i											w_1	w_2	w_3	w_3	w_6	w_7

§1.4 THE SHORTEST PATHS BETWEEN ALL PAIRS OF NODES

Turning finally to the problem of finding the shortest paths between all pairs of nodes, we recall our previous statement that a 'brute force' approach to this problem is needlessly arduous and hence unacceptable. What we seek is an 'intelligent' approach which alleviates the computational burden, since real-life networks containing 500 nodes or more are very common indeed and solving 249,500 shortest path problems is simply not a cherished prospect!

A Special Case: Undirected Networks

In the special case of undirected networks the matrix of distance $D = [d_{ij}]$ is symmetric, and hence we need evaluate only $N(N-1)/2$ entries in the distance matrix, $B = [b_{ij}]$. An approach which may vie with the Revised Cascade Method described below (a general method applicable to directed or undirected networks, with either symmetric or asymmetric distance matrices) involves a slight generalization of the Dijkestra Method described in §1.1.

The basic idea is simply the following. Since the Dijkstra Method determines a tree of shortest paths from the source node s to all other nodes, we simultaneously obtain the distance between any pair of nodes lying on a (unique) path in the tree. Consequently, although the tree has N-1 arcs, we may be able to determine more than N-1 entries in the distance matrix B based on the tree just constructed. By the judicious choice of the nodes s to serve as sources for different trees the complete B-matrix may be determined in few iterations.

To this end, recall that the base of a network relative to source s was defined to be the set of all arcs on minimal paths from s to all other nodes of the network. s is called the root of the base; and an arc is called basic if it belongs to the base. Notice that only in the case of a unique minimal path from s to every other node of the network will the base be a tree; see Fig. 1.2.

The procedure is as follows:

(i) At the start, choose s to be any node, say node 1. Construct the distance base using Dijkstra's algorithm, for example, and determine the distance, denoted by b_{ij}, between all pairs of nodes i and j lying on a path in the base.

(ii) At any subsequent iteration choose the node (row or column in B) with the least number of entries to serve as source s. Utilize previously determined b_{ij}'s to construct part of the tree with root s, and the Dijkstra Method to determine the remainder of the tree. Evaluate all the b_{ij}'s between all pairs of nodes i and j lying on a path in the tree.

(iii) Repeat Step (ii) until the matrix B is completed.

As illustration, consider the network of Fig. 1-12. It is composed of N=5 nodes and A=10 arcs. The distances between nodes are shown on the arcs. The steps of iteration are shown in Fig. 1-13. Notice that in step (i) already 8 out of the 10 desired values are determined. In both steps (ii) and (iii) only one comparison is needed to determine the corresponding entry b_{ij}.

The General Case: The Revised Cascade Method (RCM)

We define a 'triple operation' by

$$d_{ik} = \min (d_{ik}; d_{ij} + d_{jk}) \quad (1\text{-}1)$$

in which the distance d_{ik} between nodes i and k is compared to the sum $d_{ij} + d_{jk}$ for some intermediate node j, and the minimum of these two quantities replaces d_{ik} in the matrix of distances D.

Operation (1-1) is executed in the following fashion. Let j in (1-1) be successively fixed at 1,2,...,n; for each value of j, say $j = j_0$, do the operation (1-1) for each entry d_{ik} with $i \neq j_0 \neq k$.

When $j = j_0$ node j_0 is called the pivot node and the j_0th row and j_0th column the pivot row and column, respectively. Since when j_0 is the pivot, the pivot row and column are deleted from the matrix, we are left with an $(N-1) \times (N-1)$

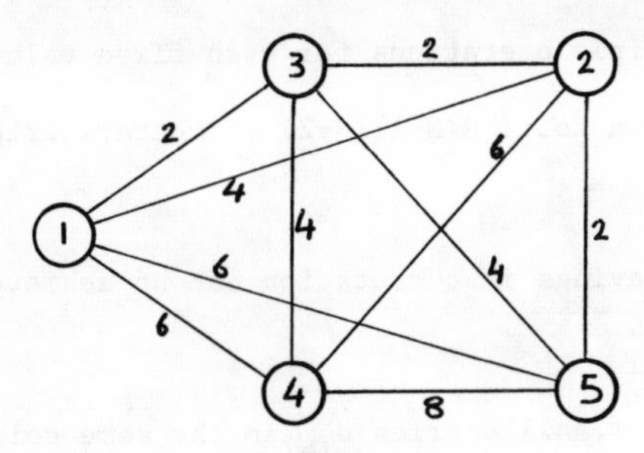

Figure 1-12

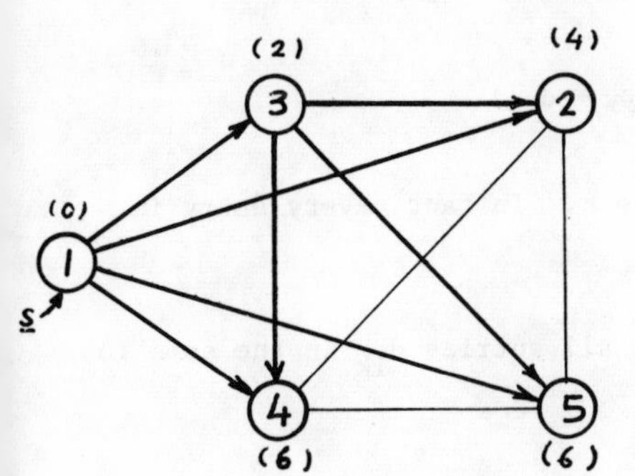

	1	2	3	4	5
1	0	4	2	6	6
2		0	2		2
3			0	4	4
4				0	
5					0

Step (i) Base yields 8 values of b_{ij}.

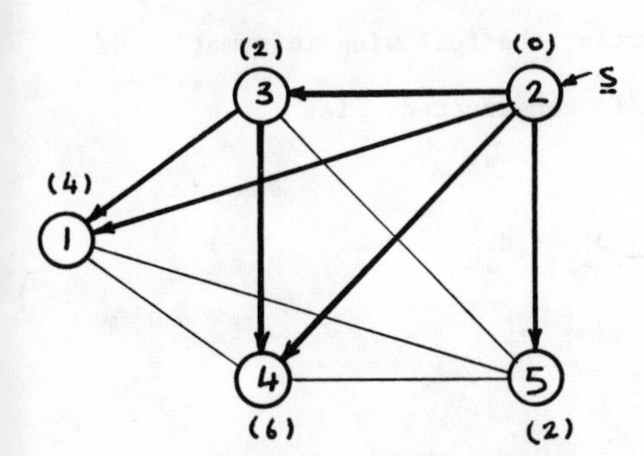

1	0				
2		0		6	
3			0		
4				0	
5					0

Step (ii) Base yields one value b_{24}; determined in one comparison.

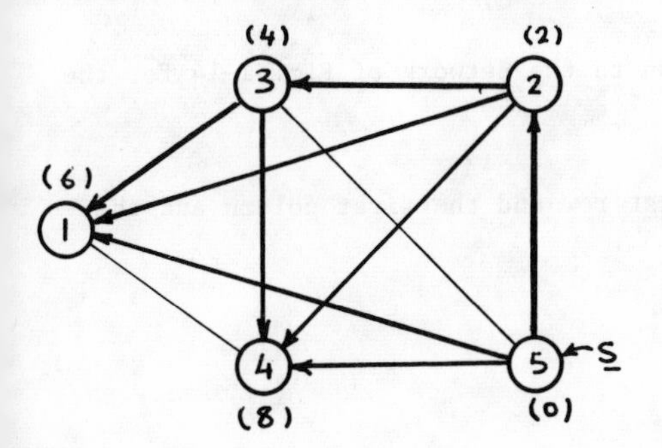

1	0				
2		0			
3			0		
4				0	8
5					0

Step (iii) Base yields one value b_{45}; determined in one comparison.

Figure 1-13

matrix of which N-1 entries (on the main diagonal) are = 0, and we need only do $(N-1)^2 - (N-1) = (N-1)(N-2)$ triple operations for each fixed value of j. As there are N values of j, there are, in total $N(N-1)(N-2)$ elementary triple operations, as was previously asserted.

In fact, some further savings in computation can be achieved, as was noted by Murchland[24]:

1. If $d_{j_0 k}$ in the pivot row = ∞, all entries d_{ik} in the same column as this $d_{j_0 k}$ remain unchanged. For example, if i = 2, k = 5 and j = 3, then in evaluating the triple operation

$$d_{25} = \min(d_{25}; d_{23} + d_{35})$$

   the minimum is clearly given by d_{25} if $d_{35} = \infty$. In fact, every entry in column 5 will remain unchanged.

2. Similarly, if d_{ij_0} in the pivot column = ∞, all entries d_{ik} in the same row as d_{ij_0} remain unchanged.

The matrix thus generated is the shortest distance matrix, as is proven in Floyd's paper. To determine the shortest routes, the following information is generated while the shortest distance matrix is constructed: let

$$r_{ik} = \begin{cases} \underline{k} & \text{if } d_{ik} \leq d_{ij} + d_{jk} \\ r_{ij} & \text{otherwise} \end{cases} \tag{1-2}$$

The matrix $R \equiv [r_{ik}]$ is the route matrix.

We demonstrate the method by application to the network of Fig. 1-14 for the cases $\underline{j} = 1$ and 2.

When the pivot $\underline{j} = 1$, we delete the first row and the first column and the resultant matrix is

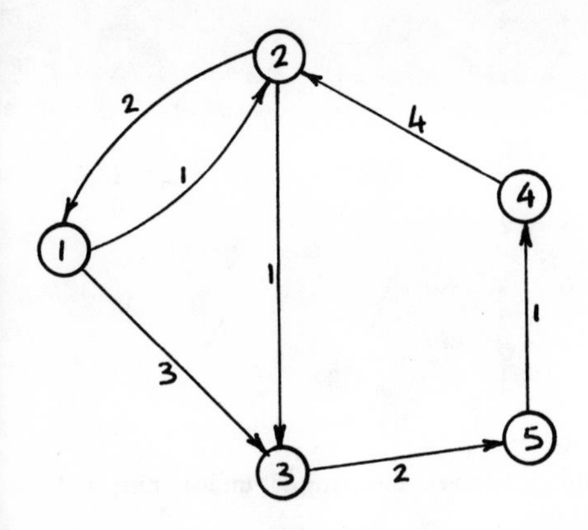

	1	2	3	4	5
1	0	1	3	∞	∞
2	2	0	1	∞	∞
3	∞	∞	0	∞	2
4	∞	4	∞	0	∞
5	∞	∞	∞	1	0

Matrix D

Figure 1-14

	1	2	3	4	5
1	0	1	2	5	4
2	2	0	1	4	3
3	9	7	0	3	2
4	6	4	5	0	7
5	7	5	6	1	0

Matrix B

	1	2	3	4	5
1	-	2	2	2	2
2	1	-	3	3	3
3	5	5	-	5	5
4	2	2	2	-	2
5	4	4	4	4	-

The route Matrix R.

Ex: $r_{52} = 4$, $r_{42} = 2$. Therefore, route from 5 to 2 is (5,4,2).

Figure 1-15

	1	2	3	↓ 4	↓ 5
1					
2		0	1	∞	∞
→ 3		∞	0	∞	2
→ 4		4	∞	0	∞
→ 5		∞	∞	1	0

Since (3,1) = (4,1) = (5,1) = ∞, rows 3, 4 and 5 remain unchanged under the triple operation (1-1). Moreover, since (1,4) = (1,5) = ∞, columns 4 and 5 also remain unchanged. Unchanged rows and columns are indicated by arrows. Therefore, the only entry to be evaluated is d_{23},

$$d_{23} = \min [d_{23};\ d_{21} + d_{13}] = d_{23} = 1; \quad r_{23} = 3.$$

Next, consider the case $\underline{j} = 2$. Deleting the pivot row and column we obtain:

	1	2	3	↓ 4	↓ 5
1	0		3	∞	∞
2					
→ 3	∞		0	∞	2
4	∞		∞	0	∞
→ 5	∞		∞	1	0

Here, rows 3 and 5 and columns 4 and 5 remain unchanged, as indicated by arrows. Consider d_{13},

$$d_{13} = \min [d_{13};\ d_{12} + d_{23}] = \min [3;\ 1+1] = 2; \quad r_{13} = r_{12}.$$

The value 2 now replaces the 3 in the matrix. We continue,

$$d_{41} = \min [d_{41};\ d_{42} + d_{21}] = \min [\infty;\ 4 + 2] = 6; \ r_{41} = r_{42}.$$

The value 6 now replaces the ∞ in the matrix. Next,

$$d_{43} = \min [d_{43};\ d_{42} + d_{23}] = \min [\infty;\ 4 + 1] = 5; \ r_{43} = r_{42}.$$

The value 5 now replaces the ∞ in the matrix. This completes the calculations for $\underline{j} = 2$. Similar calculations lead to the shortest distance matrix shown in Fig. 1-15.

§1.5 A SPECIAL CASE: DIRECTED ACYCLIC NETWORKS

The case of directed acyclic networks appears prominently in the operations research literature because of its relevance to activity networks, a subject better known under various acronyms such as PERT, CPM, PEP, and others, and which we shall encounter in greater detail in Part 4 below. It turns out that due to the special structure of such networks, to wit, that the nodes can always be numbered such that an arc leads from a small numbered node to a larger one, and hence the distance matrix D is upper triangular, all four Problems posed above (p.6) can be answered in a rather simple and straightforward manner.

Consider first the problem of finding the shortest path between nodes 1 and n, which is Problem I. Construct the matrix of distances D as indicated above, viz.,D is upper triangular. Let $\delta_1 = 0$, then for all nodes with positive entry in the first row of D put $\delta_i = d_{1i}$. At any step (node) j, consider the set of nodes (rows) {i} which connect with node j. Node j is labeled

$$\delta_j = \min_{i \to j}\{\delta_i + d_{ij}\} \tag{1-3}$$

then move to node j+1, and so on. The process is terminated when node n is labeled. δ_n is the length of the shortest path, and the path(s) itself is determined in the usual fashion by tracing backwards from n to all nodes such that $\delta_i + d_{ij} = \delta_j$, $j = n, n-1, \ldots, 1$.

As usual, solving Problem I also solves Problem III, that of finding the shortest path from 1 to all other nodes of the network. The labels $\{\delta_j\}$ give the lengths of such paths; the paths themselves are determined for each node in the usual manner.

It is interesting to note that the method of solving Problem I (i.e., the recursive application of Eq. (1-3)) also solves Problem IV, that if finding the shortest paths between all pairs of nodes. The very special structure of the matrix D indicates that there are no paths between all pairs of nodes, only possibly between a node and other nodes lying 'on its right'. Consequently, to find the shortest path from s to t, t > s, consider only the submatrix bound by rows s and t and columns s and t, and apply the same procedure outlined above with $\delta_s = 0$.

There remains only Problem II, that of determining the m-shortest paths from 1 to n. It is easily solved by a slight modification of the procedure for solving Problem I as follows.

For convenience of notation let $\min_1$ denote the minimum, $\min_2$ denote the second minimum, and so on. Then the general step of the procedure reads as follows. At any step (i.e., node) j, consider the set of nodes (rows) {i} which connect with node j. Then the kth shortest distance is obtained from

$$\delta_j^{(k)} = \min_k \{\delta_i^{(r)} + d_{ij}\} \qquad \text{for } 1 < r < k, \text{ all } i \to j, \quad k=1,\dots,m \qquad (1-$$

The process terminates when n is labeled with $\delta_n^{(k)}$, k=1,...,m.

To illustrate, consider the network of Fig. 1-16, together with its D matrix. It is immediately obvious that $(\delta_1,\dots,\delta_9) = (0,2,4,1,6,12,6,6,10)$. The distance matrix is upper triangular and is also shown in Fig. 1-16.

Suppose now that we are interested in the 3-shortest paths from 1 to 9. We proceed as follows:

node 2: $\delta_2^{(1)} = 2 = \delta_2^{(2)} = \delta_2^{(3)}$

3: $\delta_3^{(1)} = 4 = \delta_3^{(2)} = \delta_3^{(3)}$

4: $\delta_4^{(1)} = 1 = \delta_4^{(2)} = \delta_4^{(3)}$

5: $\delta_5^{(1)} = 6; \delta_5^{(2)} = 10 = \delta_5^{(3)}$

6: $\delta_6^{(1)} = 12 = \delta_6^{(2)} = \delta_6^{(3)}$

7: $\delta_7^{(1)} = 6; \delta_7^{(2)} = 13 = \delta_7^{(3)}$

8: $\delta_8^{(1)} = 6; \delta_8^{(2)} = 16; \delta_8^{(3)} = 20$

9: $\delta_9^{(1)} = 10; \delta_9^{(2)} = 13; \delta_9^{(3)} = 17$

As a sample calculation, consider node 9. Three nodes connect with it: nodes 6, 7 and 8. Then

$$\delta_9^{(1)} = \min_1 \{\delta_6^{(1)} + d_{69};\ \delta_7^{(1)} + d_{79};\ \delta_8^{(1)} + d_{89}\} = 10$$

$$\delta_9^{(2)} = \min_2 \{\delta_6^{(1)} + d_{69};\ \delta_6^{(2)} + d_{69};\ \delta_7^{(1)} + d_{79};\ \delta_7^{(2)} + d_{79};\ \delta_8^{(1)} + d_{89};\ \delta_8^{(2)} + d_{89}\} = 13$$

$$\delta_9^{(3)} = \min_3 \{\delta_6^{(1)} + d_{69};\ \delta_6^{(2)} + d_{69};\ \ldots;\ \delta_8^{(3)} + d_{89}\} = 17$$

An alternative dynamic programming formulation for this problem was proposed by Fan and Wang in 1963 in an unpublished report. The method was later incorporated as Appendix 3 of their book, ref. [12].

In large networks (some PERT networks reportedly contain 350,000 nodes and over 500,000 arcs!), maintaining the complete D matrix in the fast-access computer memory may be infeasible even for large computer installations, since it still involves storing variables of the order of n^2. Although the problem here is not as acute as it is in the general algorithm for solving Problem IV, it is still advantageous to devise a method which does not overwhelm small or medium size computers.

Fortunately, this is easy to accomplish, again due to the special structure of the D matrix. We suggest two approaches, leaving it to the reader to formalize them and fill in the necessary details, or perhaps suggest other approaches.

(1) Subdivide the D matrix by 'vertical' partitions into several sections, the first section contains columns 1 through j_1 (and, of course, rows 1 through j_1), the second section contains columns $j_1 + 1$ through j_2 (and rows j_1 through j_2) etc.; see Fig. 1-17. Naturally, these sections are chosen such that they contain approximately equal amounts of data. Calculate δ_i, $i=1,\ldots,j_1$, for which only the first section of the D matrix needs to be stored in the fast-access computer memory. Then replace the first section with the second section. Now calculate δ_i, $i=j_1 + 1,\ldots,j_2$, for which only the second section together with $\delta_1,\ldots,\delta_j$ are needed, and so on.

(2) Partition, or decompose†, the large network into a number of smaller subgraphs

† A 'decomposition' approach was suggested by Parikh and Jewell, ref. [25], which is unfortunately incorrect. A correct decomposition approach, however, can be found in ref. [20].

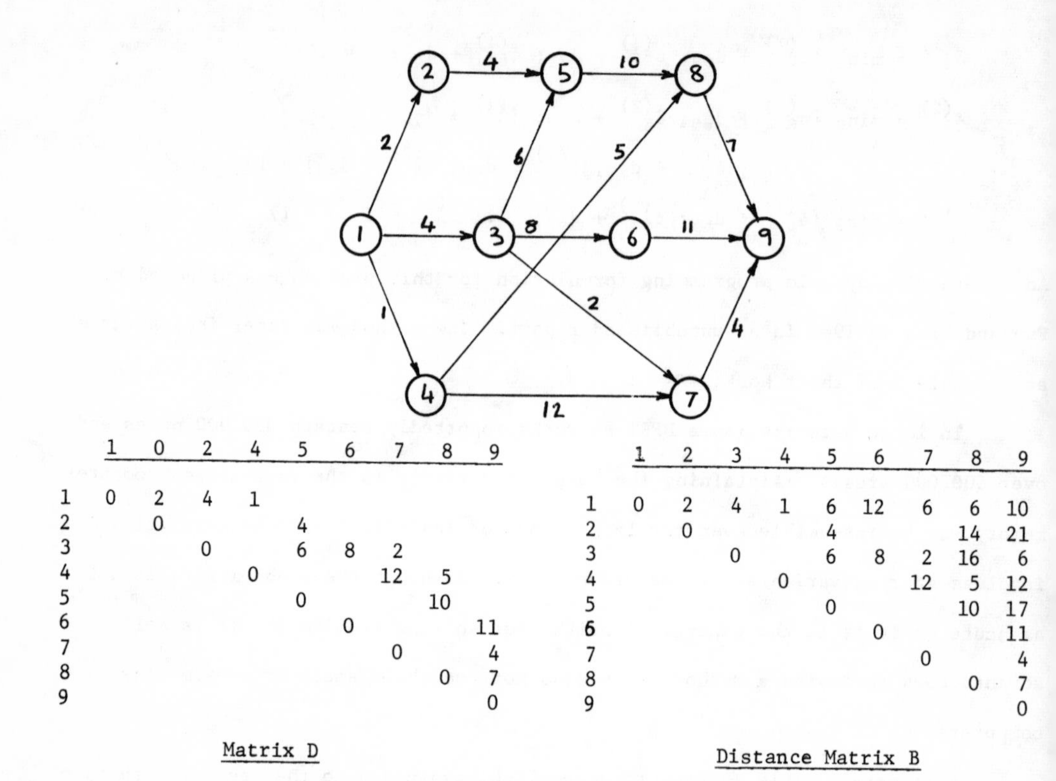

	1	0	2	4	5	6	7	8	9
1	0	2	4	1					
2		0			4				
3			0		6	8	2		
4				0			12	5	
5					0			10	
6						0			11
7							0		4
8								0	7
9									0

Matrix D

	1	2	3	4	5	6	7	8	9
1	0	2	4	1	6	12	6	6	10
2		0			4			14	21
3			0		6	8	2	16	6
4				0			12	5	12
5					0			10	17
6						0			11
7							0		4
8								0	7
9									0

Distance Matrix B

Figure 1-16

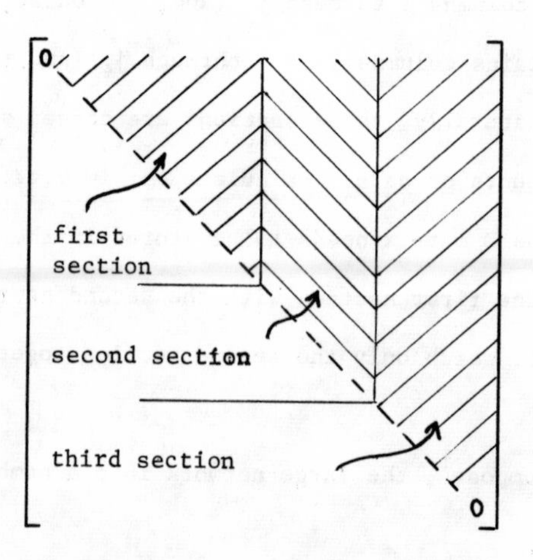

The Subdivision of the D-Matrix

Figure 1-17

such that these subgraphs are minimally connected to each other. This approach would be highly successful in the case of a network composed of few clusters. Determine the shortest path between the origin and terminal of each subgraph independently, i.e., ignoring the connecting arcs. Then join the subgraphs, two at a time, until the original network is reconstructed.

§1.6 RELATED TOPICS

The shortest path problem bears close relationship to other problems which, as originally formulated, do not show any 'shortest path' characteristics.

Relationships usually come about in one of two forms. Either the problem is re-cast, i.e., re-formulated, such that it yields a shortest path formulation, or the methodologies of the two problems are, in fact, the same except that one problem (say the shortest path problem) is a special case (or a general case) of the other.

In this section four such related problems are discussed. By necessity, the discussion is brief. For a more detailed study of these problems, and of related topics to them, the reader is advised to go to the references listed at the end of the chapter.

(i) The Most Reliable Route

Suppose the network $G \equiv (N,A)$ is a representation of a large scale system, where the arcs represent components of the system and nodes represent junction points among the components. Suppose, further, that the origin s now designates the input junction point to the system and the terminal t designates the output junction point. The input must follow one, and only one, path from s to t. The problem is to determine the most reliable path from s to t, where reliability is defined as the probability of nonfailure, and the reliability $R(\Pi)$ of a path Π composed of arcs a,b,...,w is the product of the reliabilities of the individual arcs.

This problem can be reformulated to correspond exactly to the (algebraically) shortest route problem, since we are interested in the maximal reliability. Let $d_{ij} = \log p_{ij}$, where p_{ij} is the reliability of arc (i,j). Obviously d_{ij} is ≤ 0. We have,

$$\log R = \sum_{(i,j)\in\Pi} d_{ij} ,$$

and the most reliable route in the network is the (algebraically) shortest loopless path(s) from s to t (i.e., the 'most negative' path).

If the network is directed and acyclic, this problem can be solved very easily by the methods outlined in section 1.4. If, on the other hand, the network is of the general type, the solution of this problem is easily obtained by replacing the 'triple operation' of Eq. (1-1) by

$$p_{ik} = \max (p_{ik}, p_{ij} \times p_{jk}) \tag{1-5}$$

and proceeding as before.

(ii) The Maximum Capacity Route

In this problem every arc (i,j) of the network has associated with it a capacity $c_{ij} < \infty$. The arc capacity indicates the maximum amount of flow that can pass from i to j. The problem is to find a route from s to t such that

$$\min (c_{s1}, c_{12}, \ldots, c_{nt})$$

is maximum.

If the network is undirected, it can be easily shown that the maximum capacity routes between all pairs of nodes is a maximal spanning tree; i.e., it is a tree such that for any arc (i,j) not in the tree we have

$$c_{ij} \leq \min (c_{i1}, c_{12}, \ldots, c_{nj}).$$

In other words, any arc not in the tree must be of capacity not larger than the minimal capacity of the (unique) path in the tree joining the terminal nodes of the arc.

The manner in which such a maximal spanning tree is constructed is very simple, and is due to Kruskal[21]; see also Prim[30]. Begin by selecting the arc with the largest capacity; at each successive stage, select from all arcs not previously selected the largest capacity arc that completes no cycle with previously selected arcs. After n-1 arcs have been selected a maximal spanning tree has been constructed.

For the general type of network, i.e., directed networks with loops and non-symmetric D matrices, re-define the 'triple operation' of Eq.(1-1) to

$$c_{ik} = \max\ [c_{ik};\ \min\ (c_{ij}, c_{jk})] \tag{1-6}$$

Application of the RCM using this new operation yields the desired result.

(iii) <u>The Transhipment Problem</u>

The Transhipment problem is a generalization of the well-known Transportation problem of linear programming. In the Transhipment problem, it is desired to ship a commodity which is available in quantities $a_1, \ldots, a_m$ at the m sources $\underline{s}_1, \ldots, \underline{s}_m$, respectively, to a number of terminals $\underline{t}, \ldots, \underline{t}_n$ where it is demanded in quantities $d_1, \ldots, d_n$, respectively, and <u>minimize the total cost</u> of transportation, where the cost of transporting a unit of the commodity in arc (i,j) is c_{ij}. (The name 'transhipment' stems from the fact that one need not ship directly from a source $\underline{s}_i$ to a terminal $\underline{t}_j$, but may ship <u>via</u> another node or nodes.)

In the <u>special case of one source</u> $\underline{s}$ <u>and one terminal</u> $\underline{t}$, the solution of the transhipment problem is really given by the shortest path from $\underline{s}$ to $\underline{t}$, and conversely, as was pointed out by Charnes and Cooper[5]. The general case can be solved by the methods of Chapter 2.

(iv) <u>The Traveling Salesman Problem</u>

The Traveling Salesman problem is the problem of a salesman who starts at one city $\underline{s}$ and visits each one of n-1 cities exactly once and finally returns to his starting position with minimum distance traveled. Methematically, given a finite set $\{1,2,\ldots,i,\ldots,n\}$ of nodes and the distance matrix $D = [d_{ij}]$ between every ordered pair (i,j), find the sequence of nodes $(i_1, i_2, \ldots, i_n)$ such that:

(i) Every node appears in the sequence exactly once.

(ii) The total length $\sum_{k=1}^{n-1} d_{i_k, i_{k+1}} + d_{i_n, i_1}$ of the sequence is minimal.

These two requirements imply that the sequence is a cycle. In other words: it is desired to find the shortest cycle from s and back to s which passes through all other nodes exactly once.

The simplicity of the statement of the problem is certainly misleading. Until very recently there existed no computationally feasible algorithm for its solution, albeit there exist several mathematical programming models. In fact, the practical resolution of the problem was achieved through the use of 'reliable heuristics'† of the 'branch-and-bound'[13] and "implicit enumeration"[16] variety.

What relationship (if any) is there between the Traveling Salesman problem and the shortest path problem? Actually, there is no relationship at all between the two, but there is a strong relationship between the Traveling Salesman problem and the longest path problem. In a recent paper by Hardgrave and Nemhauser[17] the following was established:

1. It is possible to reformulate a minimum cycle problem of a network G into a maximum distance problem of another network G' constructed from G in the following fashion: (i) The nodes of G' are the nodes of G plus one additional node, denoted by s'; (ii) The arcs of G' are the arcs of G except that all arrows going into s are replaced by arrows going into s'; (iii) the distances d'_{ij} of G' are defined by

$$d'_{ij} = \begin{cases} 0 & \text{if } i=j \\ K-d_{ij} & \text{if } i\neq j \text{ and } j\neq s' \\ K-d_{is} & \text{if } i\neq j \text{ and } j=s' \end{cases}$$

and K is a constant strictly greater than σ, σ being the sum of the n longest d_{ij}'s in G.

---

† The terms 'reliable heuristics' were coined by Pierce and Hatfield[26] and signify search procedures which, though based on heuristic rules, are 'reliable' in the sense that if allowed to run to completion they guarantee to yield the optimal solution or indicate infeasibility [16].

2. It is possible to reformulate a minimum tour[†] problem of a network H into a minimum cycle problem of a network G which is constructed from H in the following fashion: (i) The nodes of G are the nodes of H; (ii) In G, nodes _i_ and _j_ are joined by a directed arc (i,j) if, and only if, there is a directed path in H from _i_ to _j_; (iii) The lengths d_{ij} of G are the lengths of the shortest path in H between _i_ and _j_.

Consequently, we can go from a tour formulation (which is a meaningful problem in its own right and has received little, if any, attention) to a cycle formulation to a longest path formulation. A computationally good solution to the latter problem provides a solution to the traveling salesman problem. Such a solution is still lacking.

As a final comment, a natural extension of the Traveling Salesman problem is the problem of m salesmen who wish to divide the territory of n cities among themselves. This, and extensions thereof, are the subject of active research today. (See the survey paper of Bellmore and Nemhauser[3] and the letter to the editor by Akkanad and Turban[1].)

---

† In the definition of a tour, condition (i) above is relaxed to: "(i) Every node appears in the sequence at least once." Hence, the salesman may visit the same city more than once before returning to his starting position.

REFERENCES TO CHAPTER 1

[1] Akkanad, M. I. and E. Turban, "Some Comments on the Traveling Salesman Problem", Oper. Res., Vol. 17, No. 3, May-June 1969.

[2] Bellman, R. and R. Kalaba, "On the kth Best Policies", Journal of SIAM, Vol. 8, No. 4, Dec. 1960.

[3] Bellmore, M. and G. L. Nemhauser, "The Traveling Salesman Problem: A Survey", Oper. Res., Vol. 16, No. 3, May-June 1968.

[4] Busacker, R. G. and T. L. Saaty, "Finite Graphs and Networks", McGraw-Hill, 1965.

[5] Charnes, A. and W. W. Cooper, "A Network Interpretation and a Directed Subdual Algorithm for Critical Path Scheduling", Jour. of Ind. Eng., Vol. 13, No. 4, 1962.

[6] Clarke, S., A. Kirkonian and J. Rausen, "Computing the n Best Loopless Paths in a Network", Journal of SIAM, Vol. II, No. 4, Dec. 1963,pp.1096-1102.

[7] Dantzig, G. B., "On the Shortest Route Through a Network", Mgt. Sc., Vol. 6, No. 2, Jan. 1960, pp. 187-190.

[8] ____________, "Linear Programming and Extensions", Princeton Univ. Press, 1963.

[9] Dijkstra, E. W., "A Note on Two Problems in Connexion with Graphs", Numerische Mathematik, Vol. 1, pp. 269-271, 1959.

[10] Dreyfus, S. E., "An Appraisal of Some Shortest-Path Algorithms", Oper. Res., Vol. 17, No. 3, May-June 1969.

[11] Elmaghraby, S. E., "The Design of Production Systems", Reinhold, 1966, pp. 98-140.

[12] Fan, L. T. and C. W. Wang, "The Discrete Maximum Principle", Wiley, 1964.

[13] Farbey, B. A., A. H. Land and J. D. Murchland, "The Cascade Algorithm for Finding the Minimum Distances on a Graph", Transport Network Theory Unit, London School of Economics, Nov. 1964.

[14] Floyd, R. W., "Algorithm 97, Shortest Path", Comm. ACM, Vol. 5, p.345, 1962.

[15] Ford, L. R. and D. R. Fulkerson, "Flows in Networks", Princeton Univ. Press, 1962.

[16] Geoffrion, A. M., "An Improved Implicit Enumeration Approach for Integer Programming", Oper. Res., Vol. 17, No. 3, May-June 1969.

[17] Hardgrave, W. W. and G. L. Nemhauser, "On the Relation Between the Traveling Salesman and the Longest Path Problems", Oper. Res., Vol. 10, No. 5, Sept.-Oct. 1962.

[18] Hoffman, W. and R. Pavley, "A Method for the Solution of the nth Best Path Problem", Jour. Assoc. Comput. Mach., Vol. 6, 1959, pp. 506-514.

[19] Hu, T. C., "Revised Matrix Algorithms for Shortest Paths", IBM Research Center, Research Paper RC-1478, Sept. 1965.

[20] ________ , "A Decomposition Algorithm for Shortest Paths in a Network", IBM Watson Research Center, Yorktown Heights, N. Y., Feb. 1966.

[21] Kruskal, J. B., Jr., "On the Shortest Spanning Subtree of a Graph and the Traveling Salesman Problem", Proc. Amer. Math. Soc., Vol. 7, 1956, pp. 48-50.

[22] Minty, G. J., "A Comment on the Shortest-Route Problem", Oper. Res. Vol. 5, No. 5, Oct. 1957, p. 724.

[23] Moore, E. F., "The Shortest Path Through a Maze", Proc. Intl. Symp. on The Theory of Switching, Part II, April 2-5, 1957. The Annals of the Computation Laboratory of Harvard Univ., Vol. 30, Harvard Univ. Press, 1959.

[24] Murchland, J. D., "A New Method for Finding All Elementary Paths in a Complete Directed Graph", LSE-TNT-22, London School of Economics, Oct. 1965.

[25] Parikh, S. C., and W. S. Jewell, "Decomposition of Project Networks", Mgt. Sc., Vol. 11, No. 3, Jan. 1965, pp. 438-443.

[26] Pierce, J. F., Jr. and D. J. Hatfield, "Production Sequencing by Combinatorial Programming,"Ch. 17 of Operations Research and Management Information Systems, Ed. by J. F. Pierce, Jr., TAPPI, 1966.

[27] Pollack, M., "Solutions of the kth Best Route Through a Network - A Review", Journ. Math. Analysis and Appl., Vol. 3, No. 3, Dec. 1961, pp. 547-659.

[28] __________, "The kth Best Route Through a Network", Oper. Res.,Vol. 9, No. 4, July-August 1961, pp. 578-580.

[29] __________and W. Wiebenson, "Solutions of the Shortest Route Problem-A Review", Oper. Res., Vol. 8, No. 2, March-April 1960, pp. 224-230.

[30] Prim, R. C., "Shortest Connection Networks and Some Generalizations", Bell System Technical Jour., Vol. 36, 1957, pp. 1389-1401.

CHAPTER 2

NETWORKS OF FLOW

Contents Page

§2.1 INTRODUCTION

A natural, though by no means elementary, question that comes to mind when studying networks is that of the maximum possible flow between two specified nodes of the network. The physical problem arises in almost every instance in which commodities - physical or otherwise - flow from a source s to a destination t. Thus, in a traffic network we may be interested in the maximum rate of traffic flow between two cities; in an electric power distribution network we may be interested in the maximum power transmitted from a generating station (or stations) to a particular location; in a natural gas distribution network we may be interested in the maximum rate at which gas can be supplied to a particular consumer, etc.

This is the most basic problem in network flows, whose solution is readily available through a 'labeling procedure' due to Ford and Fulkerson[5].

There are two important extensions to the one-source to one-terminal flow problem.

The first is the 'multi-terminal' flow problem, in which several nodes are designated as (s,t)-pairs, with the same commodity flowing through the network. For example, this is the case of a telephone network in which any unordered pair of the n cities covered by the network may indeed serve as the (s,t)-pair.

One's first impulse when confronted with the problem of determining the maximum flow between all ordered pairs of nodes in a multi-terminal flow network is to repeat the procedure for the single (s,t)-pair as many times as is needed. A little reflection would indicate that in an undirected network of n nodes one would perform $n(n-1)/2$ such optimum evaluations. If n is large, which is usually the case, the amount of work involved can certainly be staggering. Fortunately, this is not necessary. As we shall see below, a most elegant procedure due to Gomory and Hu[8] requires the solution of only n-1 maximum flow problems and, what is more pleasing, these problems usually get successively smaller in size as calculation proceeds.

The second is the 'multi-commodity' flow problem, in which several commodities flow simultaneously from designated sources to designated terminals. The sources and the terminals may be different for different commodities. Two questions are usually asked: 1) to maximize the total sum of the different flows; this is called the 'maximum multi-commodity flow problem'; 2) to prescribe lower bounds on each of the flow values and ask if it is feasible; this is called the 'feasibility problem'.†

The solution of the multi-commodity flow problem involves some delicate arguments of duality which, once understood, and mastered, lead immediately to an elegant procedure requiring no more than the determination of the shortest path in the network at each iteration. The 'length' of the arcs at each iteration turns out to be the dual variables of the corresponding linear program. Thus, shortest path algorithms are used as subroutines in such problems; but we emphasize that the procedure is quite different from any of the procedures cited in this manuscript. We shall not discuss this problem beyond this brief mention. However,

† A feasibility problem has also been formulated for the single commodity flow network, of which this is a generalization.

s an interesting footnote, we remark that the solution to the multi-commodity roblem proposed by Ford and Fulkerson[4] in 1958 seems to have been the spark hat led, later on, to the decomposition principle in linear programming of Dantzig nd Wolfe[2].

Generally speaking, there are two kinds of flow problems, which present a lassification different from that discussed above.

First, one may be interested in maximizing the flow. Clearly, the problem of naximum flow (whether in single-or multi-terminal networks; in single commodity or multi-commodity flow) is meaningful only if the arcs of the network, or a key subset thereof, possess upper limits, called the capacities of the arcs and designated by $c(i,j)$ or c_{ij}, which no flow can exceed in that arc in the direction $i \to j$. Such networks are called, naturally enough, "capacitated networks".

Second, one may be interested in maximizing the value of the flow. In other words, if the network has A arcs and we let the vector $x = (x_{ij})$, $(i,j) \in A$, denote the vector of flow, then there is a function $g(x)$ which maps such flow into the real line and gives the value of the flow. Examples of such functions g are: the cost of flow $g(x) = p \cdot x$, a linear function, or $g(x) = a\,\delta(x) + px$, where $\delta(x) = 0$ if $x = 0$ and $= 1$ if $x > 0$, a concave function, etc. In such a case the arcs of the network may, or may not, be capacitated depending on the context of the problem.

Once the concept of flow networks is visualized, the questions related to flow that can be legitimately asked seem to multiply abundantly, as if Pandora's box has been opened. Even more interesting, many physical non-flow problems can be interpreted in flow terms to great advantage with respect to both insight into the behavior of the system and ease of computation. For instance, the following problems, among many others, have been given flow-network representation:

1. The Personnel Assignment Problem: n graduates have been hired to fill n vacant jobs. Aptitude tests, college grades and recommendations of professors help assign a 'proficiency index' a_{ij} to candidate i on job j. What is the assignment (of all n candidates) that maximizes the total proficiency score over all jobs?

2. A Plant Location-Allocation Problem: A producer wishes to manufacture a new product in a number of plants to be built specifically for the purpose. There shall be no more than m such plants; and m locations have already been selecte as the most suitable sites. There are n markets (or demand centers); the dema r_j for the jth market is known over the planning horizon of L years. How many plants should be built, and what is the allocation of markets to plants, which maximize the total discounted net revenue over the production and distribution costs?
3. A Production Inventory Problem: The market requirements $\{r_t\}$ for a particular product over a finite planning horizon are known, $t = 1,2,\ldots,T$. If x_t units are produced in period t, the cost of production is $p_t(x_t)$, and if I_t units are held in inventory at the end of period t, a holding cost equal to $h_t(I_t)$ is incurred, where $I_t = I_0 + \sum_{i=1}^{t} (x_i - r_i)$. All demand must be satisfied in the period of its occurrence. What is the production schedule (i.e., the x_i's) which maximizes the overall gain?
4. The Warehousing Problem: A wholesaler purchases stores, and sells in each of T successive periods, some commodity that is subject to known fluctuations in purchasing costs and selling prices. The wholesaler has a warehouse of fixed capacity in which new purchases and hold-overs from the previous period are stored before selling. What is his strategy of buying, storing and selling which maximizes his profit?

Within recent years the theory of flow-networks has emerged as a most powerful tool of analysis for many operations research and management science problems, as attested to by the rather large number of articles and books written on the subject Fortunately, three bibliographies are available for the interested reader: the excellent book by Ford and Fulkerson[5] and the review papers by Fulkerson[7] and Hu[10]. The remainder of this chapter may be considered as an introduction to the subject with some illustrations of its applications. In particular, the four problems mentioned above will be discussed in greater detail for the cases of linear and concave objective functions.

In order to be able to resolve these, and other problems we introduce first the basic structure on which flow theory, and especially linear flow theory, is based. As will become evident to the reader, the duality theory of linear programming, particularly the complementary slackness theorem, plays a fundamental role in the development of efficient algorithm on which the flow solutions are based. (An excellent account of duality and complementary slackness can be found in the book by Dantzig[1], Ch. 6.)

§2.2 MAXIMIZING THE (s,t)-FLOW: THE LABELING PROCEDURE

Consider a network that possesses one source s and one terminal t; this is no restriction on the generality of the approach since we can always add a "master source" and a "master terminal" to any network of multiple sources and destinations. Such a network is shown in Fig. 2.1. For the sake of concreteness of exposition,

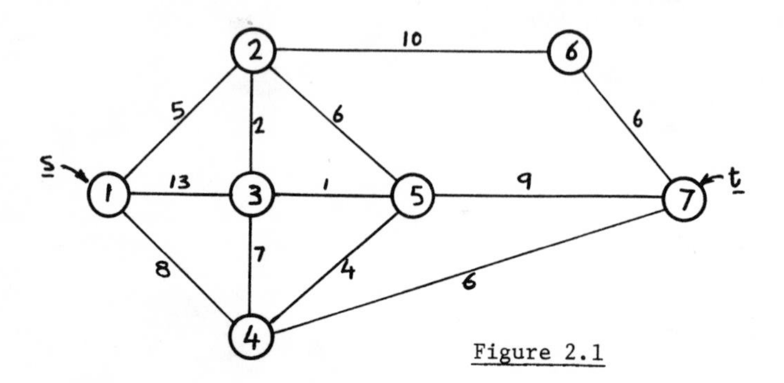

Figure 2.1

let the capacities of the various arcs be as shown: e.g., c(s,2) = 5; c(2,5) = 6, etc. Notice that the network is undirected. The problem is to determine the maximal flow between s and t, assuming infinite availability at s.

We shall determine the maximal flow between s and t by the labeling procedure due to Ford and Fulkerson[5]. In spite of the simplicity of the procedure, its verbal description is rather awkward and, unfortunately, lengthy.

To start the procedure, assume any reasonable flow $\{f(i,j)\} \geq 0$. There is no question of feasibility, since f(i,j) = 0 for all (i,j) is certainly feasible. But one can usually do better than that. In this example, we start with the flow shown in Fig. 2.2. Each arc carries two numbers (f,r), where f is the flow through the arc, $f(i,j) \leq c(i,j)$, and r is the residual capacity $r(i,j) = c(i,j) - f(i,j) \geq 0$. Notice that branches (s,2), (s,4), (3,5), and (4,t) are saturated, since r = 0 in all of them.

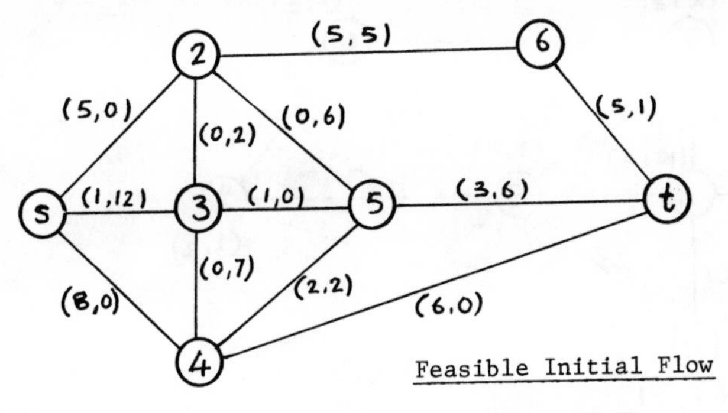

Feasible Initial Flow

Figure 2.2

Inspection of Fig. 2.2 reveals that 12 more units can be transferred from (which is assumed of infinite availability) to 3 because of the available esidual capacity in arc (s,3). This fact is recorded by labeling node 3 with 12;s): the first entry gives the quantity that can be transmitted and the second ntry the origin of such flow.

When 12 units are available at node 3, it is obvious that two units and no ore, due to the capacity of branch (3,2), can be transmitted to node 2 and seven nits to node 4. Thus, nodes 2 and 4 can be labeled with (2;3) and (7;3) respecively. Thus, in labeling node j from node 3, j = 2 or 4, the flow is the minimum of two numbers: q_3, the quantity made available at node 3 by previously determined abeling, and r(3,j), the residual capacity in branch (3,j). In general, if node j connects with node i, then node j can be labeled as $(q_j, \underline{i})$, where

$$q_j = \min\,[q_i; r(i,j)] \qquad \text{if } r(i,j) > 0. \tag{2-1}$$

Naturally, if r(i,j) = 0, node j cannot be labeled from node i. Proceeding in this fashion, the complete labeling of Fig. 2.2 is accomplished, as shown in Fig. 2.3. It is advisable to proceed in a systematic way such as: at stage k always start with the node bearing the smallest (or largest) number, label all nodes that connect with it which can be labeled and have not been labeled before (following the rule of Eq. 2-1), then proceed to the next higher (or lower)

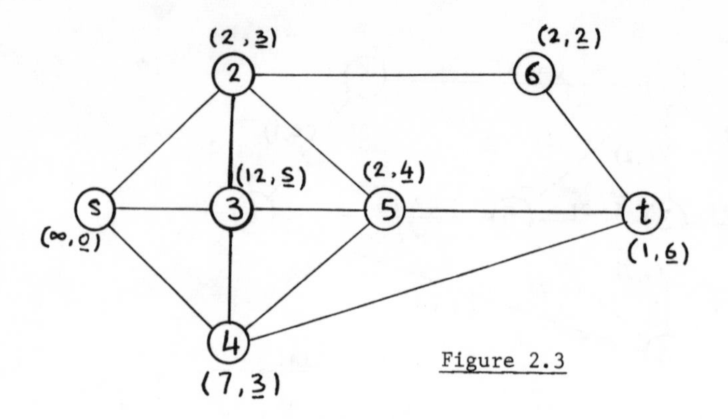

Figure 2.3

numbered node and repeat the process until all labeled nodes in stage k are exhausted. Then move to stage k + 1, i.e., to the nodes that have just been labeled from stage k. Once a node is labeled, it is not considered again for labeling. One of two conditions must be obtained:

1. The terminal t is labeled.
2. The terminal t cannot be labeled.

When the terminal node t is labeled, which is termed a "breakthrough," the flow can be increased by the amount q_t. Tracing backwards from t we can determine the chain that leads to such extra flow. The flow and residual capacities are then adjusted to reflect the increased flow, q_t, along all the branches of the chain.

The cycle of labeling and increasing the flow continues until no breakthrough is possible, i.e., until t cannot be labeled. This stage is reached in Fig. 2.4c: nodes 3 and 4 are labeled but no other node can be labeled. The maximum flow has been achieved and is equal to the sum of the flow in all branches incident on either s or t; it is equal to 18 units.

The reader should consider Fig. 2.4c for a minute. Node 5 cannot be labeled in spite of the availability of an infinite supply at s, of 8 units at node 3 and 5 units at node 4, because, in a manner of speaking, all "roads are blocked" thereafter. That is, the residual capacities are all equal to zero for all arcs leading from these three nodes. It is an interesting observation that the capacities of these "blocked" branches add up to exactly 18 units, the total flow between s and t.

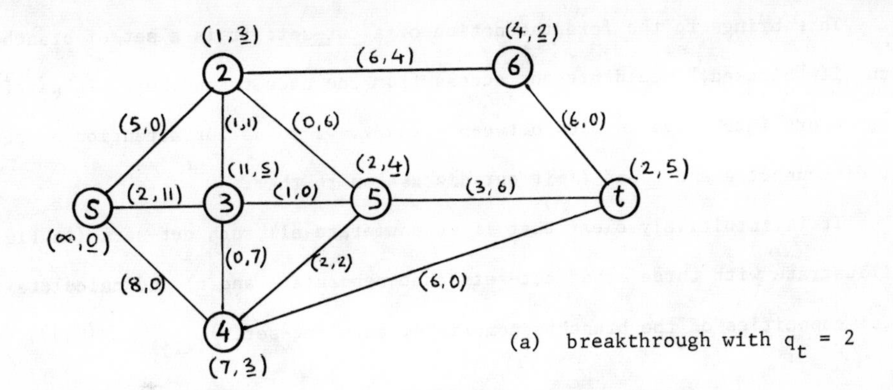

(a) breakthrough with $q_t = 2$

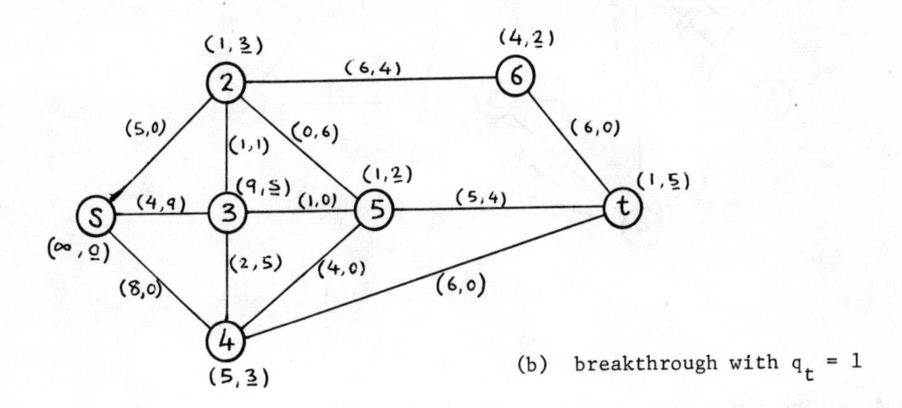

(b) breakthrough with $q_t = 1$

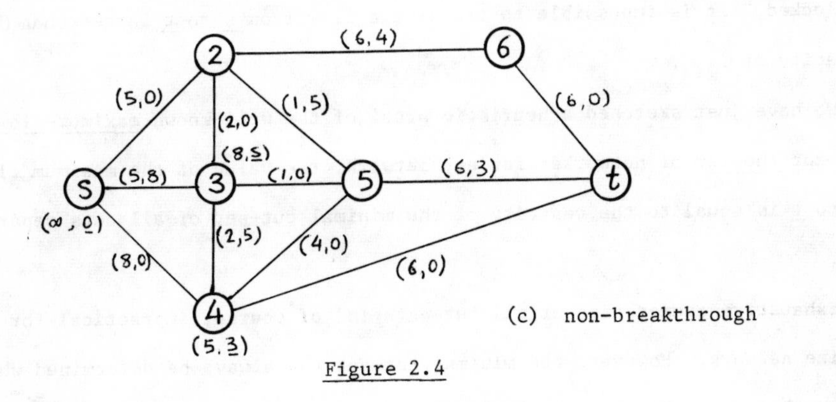

(c) non-breakthrough

Figure 2.4

This brings to the fore the notion of a cut-set: it is a set of branches which, if "blocked," would prevent access from one node to another. In particula since we are interested in flow between s and t, we focus our attention on cut-se that disconnect s and t and limit our discussion to them.

It is intuitively clear that if we enumerate all such cut-sets (in Fig. 2. we illustrate with three other cut-sets that separate s and t) and calculate the sum of capacities of the branches comprising each cut-set, $\sum_{(i,j)\varepsilon C} c(i,j)$, we

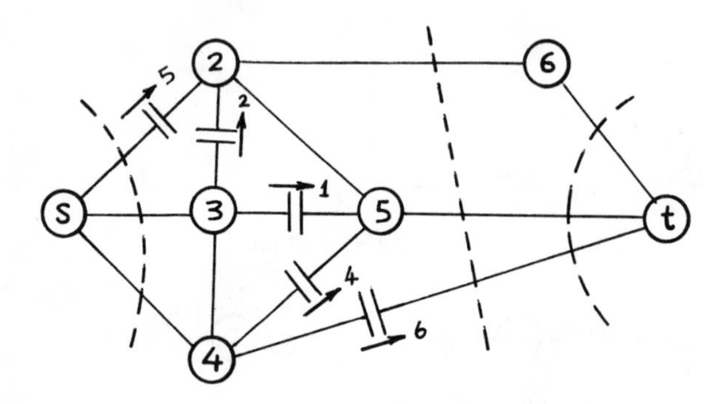

Figure 2.4d

can determine the cut-set C_{min} with the minimum capacity, denoted by C_{min} (there may be more than one such minimal cut-set). Since the minimal cut-set C_{min}, like any other cut-set, would completely obstruct all flow from s to t if its branches were "blocked," it is impossible to transmit a flow from s to t larger than C_{min}, the capacity of C_{min}.

We have just sketched a heuristic proof of the well-known maximum-flow minimum-cut theorem of networks: for any network, the value of the maximum flow from s to t is equal to the capacity of the minimal cut-set of all cuts separatin s and t.

Exhaustive enumeration of all cut-sets is, of course, impractical for any large size network. However, the minimal cut-set can always be determined when nc breakthrough is possible by restricting attention to those branches whose residual capacities equal zero. Conversely, if at any stage of the labeling process a cut-set can be discerned whose capacity is equal to the flow into t, the optimum is achieved. This is a useful check when no breakthrough is possible.

In Fig. 2.4c, such a minimal cut-set can immediately be discerned: C_{min} = {(4;t); (4;5); (3;5); (3;2); (s;2)}; and its capacity is 18 units. For illustration Fig. 2.4c has been drawn in (d) with the branches of C_{min} actually cut to demonstrate the separation of s and t and the value of C_{min}.

The labeling procedure is the tool that is used over and over again in almost all flow problems, because at one stage or the other of the solution one is usually confronted with the need to maximize the flow. For the sake of clarity we have described the procedure over the diagram of the network. It is possible, as it must be evident to the reader, to carry out the labeling and flow augmentation cycle over the matrix representation of the network. This is, in fact, how such computations are carried out in a computer.

§2.3 THE MAXIMAL FLOW IN MULTI-TERMINAL NETWORKS

In multi-terminal flow networks, we assume that the network is undirected and possesses a symmetric capacity matrix $[c(i,j)]$. It is desired to determine the maximum flow between all source-terminal pairs, and there are exactly $\binom{n}{2} = n(n-1)/2$ such pairs (because the maximal flow matrix is also symmetric).

As was indicated in §2.1, an elegant procedure, due to Gomory and Hu[8], exists which requires the solution of only n-1 maximal flow problems. The procedure really rests on two fundamental, though essentially simple, ideas.

As a way of introduction, the reader will recall our definition of a <u>spanning tree</u> of a connected undirected network G as a connected subgraph of G which contains the same nodes as G but contains no loops. It is easy to prove, e.g., by induction, that a tree contains exactly n-1 arcs, where n is the number of nodes of G.

In general, a graph, connected or not, without cycles, is called a <u>forest</u>; each connected piece of a forest is clearly a tree. Finally, for any source-terminal pair (s,t), let v(s,t) denote the value of the maximal flow between <u>s</u> and <u>t</u>. Such maximal flow can be determined by the approach of the previous section.

Now we are in a position to state the two fundamental concepts mentioned above.

The first is a form of triangular inequality:

$$v(s,t) \geq \min\,[v(s,x),\ v(x,t)];\quad s,x,t \in N \tag{2-2}$$

The proof of this inequality relies on the maximum-flow minimum-cut theorem of §2.2 For, in determining maximum flow between s and t we must have obtained a cut-set $C(X,\bar{X})$, with $s \in X$ and $t \in \bar{X}$, such that $c(X,\bar{X}) = v(s,t)$. Clearly, if $x \in X$, then $v(x,t)$ must be $\leq c(X,\bar{X})$; on the other hand, if $x \in \bar{X}$, then $v(s,x)$ must be $\leq c(X,\bar{X})$; and the inequality (2-2) follows. Simple as this inequality may seem, its consequences are far reaching. By simple induction we must have that

$$v(s,t) \geq \min [v(s,x_1), v(x_1,x_2), \ldots, v(x_r,t)] \tag{2-3}$$

where $x_1, x_2, \ldots, x_r$ are any sequence of nodes in N. Furthermore, if the network consists only of three nodes, s, x, and t and only three arcs, then applying (2-2) to each 'side of the triangle' shows that, among the three maximum flow values appearing in (2-2), two must be equal and the third no smaller than their common value. As a further consequence of (2-2) (or (2-3)) it must be true that in a network of n nodes, the function v can have at most n-1 numerically distinct values.

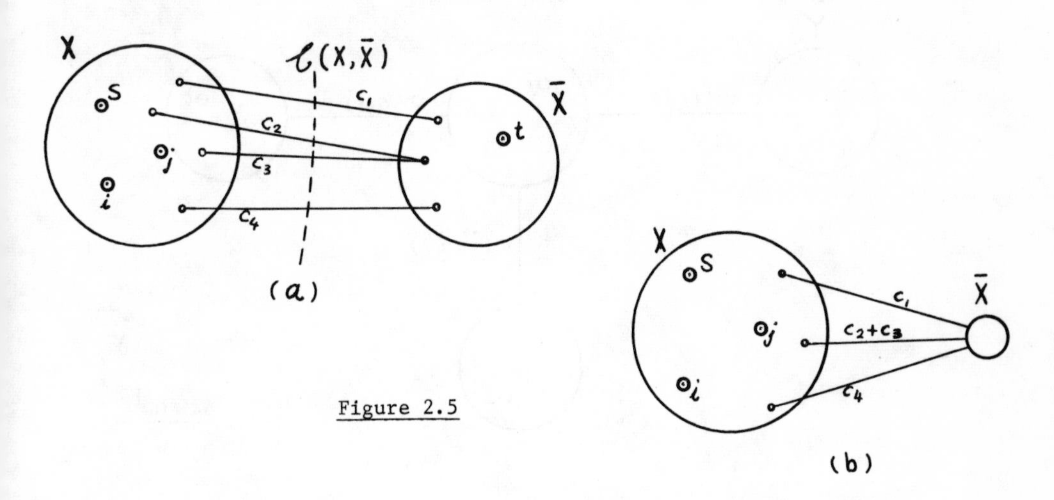

Figure 2.5

The second fundamental concept is a 'condensation property'. Let us suppose that with <u>s</u> as source and <u>t</u> as sink a maximal flow problem has been solved, thereby locating a minimal cut $C(X,\bar{X})$ with $s\varepsilon X$ and $t\varepsilon\bar{X}$, see Fig. 2.5a. Suppose now that we wish to find v(i,j) where both <u>i</u> and <u>j</u> are on the same side of $C(X,\bar{X})$; say both are in X, as shown in Fig. 2.5a. Then, for this purpose, <u>all nodes of $\bar{X}$ can be 'condensed' into a single node</u> to which all the arcs of the minimal cut are attached. The resulting network is a condensed network, as shown in Fig. 2.5b. In a sense, one may think of the condensed network as having accorded an infinite capacity to the arcs joining all pairs of nodes of $\bar{X}$.

The statement is certainly plausible. The formal proof involves the demonstration that all the nodes of $\bar{X}$ must lie on one side of the cut-set between <u>i</u> and <u>j</u>. Consequently, the condensation of all the nodes in $\bar{X}$ cannot affect the value

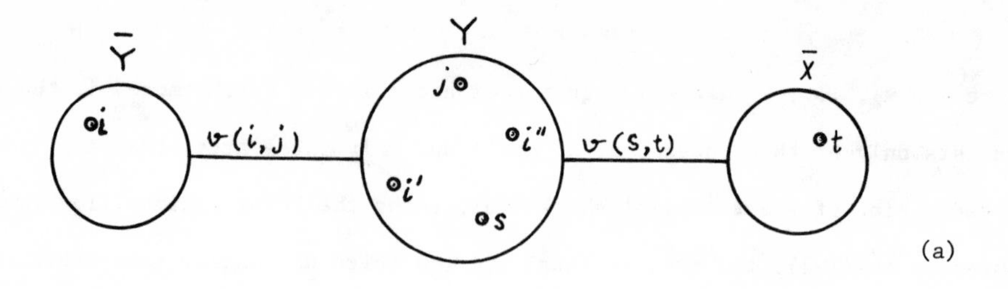

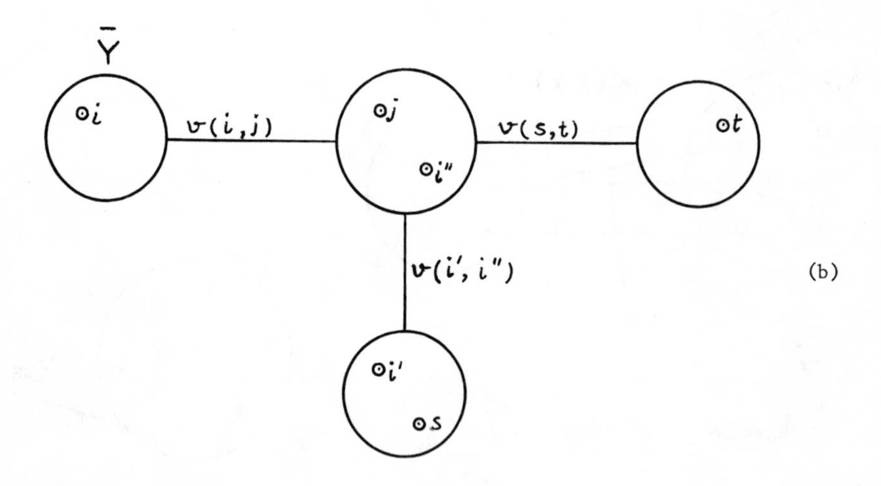

Figure 2.6

of the maximum flow from i to j.

The Gomory-Hu construction exploits this second property to the extreme. Suppose that after choosing i, $j \varepsilon X$ as the new source-terminal pair (i or j may be s), the maximal flow $v(i,j)$ and the minimal cut-set $C(Y,\bar{Y})$ between i and j are determined by the standard labeling procedure with all the nodes of the subset $\bar{X}$ condensed into one node. Notice that $\bar{X}$ is on one side of the new cut-set, see Fig. 2.6a. Next choose two nodes i', i" ε Y, say; condense all the nodes in $\bar{X}$ and $\bar{Y}$ to one node each and repeat the maximal flow determination. Continue until each condensed subset contains only one node. Obviously, this requires exactly n-1 maximal flow calculations. Moreover, the resultant is a tree which is called the cut-tree R.

As an illustration of the procedure, consider the network of Fig. 2.7 in which the capacities are shown on the arcs. The steps of iteration, the minimal cut-set at each iteration, and the final cut-tree R are shown in Fig. 2.8, steps (a) to (f).

Notice that each arc of R represents a cut-set in G, and the number $v(\tau)$ attached to the τth arc of the cut-tree is the capacity of the corresponding cut-set in G.

The remarkable conclusion is that this cut-tree gives the maximal flow between any pair of nodes. In particular, if $i,j \varepsilon N$, then

$$v(i,j) = \min [v(i,r_1), v(r_1, r_2), \ldots, v(r_k,j)] \tag{2-4}$$

where (i,r_1), (r_1,r_2), ..., (r_k,j) are the arcs of the unique path between i and j in the cut-tree R.

To see this, we must first show that each link τ of the cut-tree is indeed representing maximal flow between its two end nodes in R. We already know that the link τ represents some cut-set, i.e., maximal flow, between two nodes s and t in G. We must prove the stronger statement. This is easily accomplished by induction on the stages of iteration. In the first stage we did obtain a link τ_1,

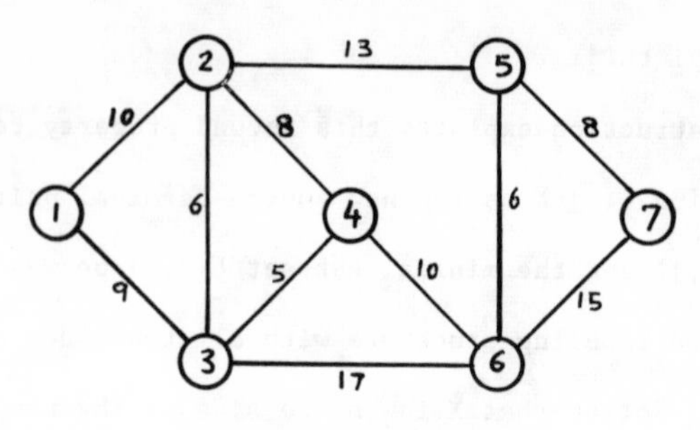

Figure 2.7

Iteration	(s-t) Nodes	Max. Flow	Cut-Set	Cut-Tree
1	(1,7)	19		
2	(2,7)	23		
3	(2,6)	36		
4	(3,6)	37		
5	(4,6)	23		
6	(5,6)	27		

Figure 2.8

say, which represents the maximal flow between the two nodes $\underline{s},\underline{t}$ chosen as the source-terminal pair. We also obtained the two subsets X and $\bar{X}$, with, say, $\underline{s}\varepsilon X$ and $\underline{t}\varepsilon\bar{X}$, see Fig. 2.5. Consider the node X to be split, with $\bar{X}$ attached to it by a link of capacity $v_1 = v(s,t)$. Let $\underline{i}$ and $\underline{j}$ be the two nodes of X for the next maximal flow problem ($\underline{i}$ or $\underline{j}$ may be $\underline{s}$). The set X divides into Y and $\bar{Y}$ with $\underline{i}\varepsilon\bar{Y}$ and $\underline{j}\varepsilon Y$, say, and suppose $\bar{X}$ is attached to Y as in Fig. 2.6a. The link between Y and $\bar{Y}$, which we shall call τ_2 is of capacity $v_2 = v(i,j)$, by construction. In other words, $\underline{i}$ and $\underline{j}$ provide the two nodes which are separated by τ_2. Of course $\underline{s}$ could be either in Y or in $\bar{Y}$ (in Fig. 2.6a it is shown in Y). If $\underline{s}\varepsilon Y$ then $\underline{s}$ and $\underline{t}$ still provide the two nodes separated by τ_1. If $\underline{s}\varepsilon\bar{Y}$, we reason as follows. Since $\underline{s}$ is now separated from $\underline{t}$ by two cut-sets, namely $C(Y,\bar{Y})$ and $C(Y,\bar{X})$, it must be true that $v_2 = c(Y,\bar{Y}) \geq c(Y,\bar{X}) = v_1$ since v_1 was the maximal flow between $\underline{s}$ and $\underline{t}$. By the triangular inequality,

$$\begin{aligned} v(j,t) &\geq \min\,[v(j,i);\ v(i,s);\ v(s,t)] \\ &= \min\,[v(i,j),\ \infty,\ v(s,t)] && \text{because } \underline{i},\underline{s}\varepsilon\bar{Y} \\ &= \min\,[v_2,v_1] \\ &= v_1 && \text{because } v_2 \geq v_1 \end{aligned} \tag{2-5}$$

But since $\underline{j}$ and $\underline{t}$ are separated by the cut-set $C(X,\bar{X})$ of capacity $v(X,\bar{X}) = v_1$, it must be true that

$$v(j,t) \leq v_1 \tag{2-6}$$

Combining the two inequalities (2-5) and (2-6) we obtain that

$$v(j,t) = v_1 \ .$$

This means that now $\underline{j}$ and $\underline{t}$ provide the two nodes separated by τ_1. Continuing in this fashion until each subset X_1, X_2, X_3, X_4, etc., contains only one node it is seen that each τ_r represents the maximal flow between its end nodes in R.

Applying (2-3) we obtain

$$v(i,j) \geq \min [v(i,r_1), v(r_1,r_2), \ldots, v(r_k,j)] \tag{2-7}$$

where (i,r_1), (r_1,r_2), ..., (r_k,j) are the arcs of the unique path between $\underline{i}$ and $\underline{j}$ in R.

The reverse inequality follows trivially from the construction of the tree: if cut-sets $\tau_1, \tau_2, \ldots, \tau_k$ separate $\underline{i}$ and $\underline{j}$, then obviously

$$v(i,j) \leq \min [v(i,r_1), v(r_1,v_2), \ldots, v(r_k,j)] \tag{2-8}$$

Inequalities (2-7) and (2-8) establish the equality of (2-4), as was to be shown.

§2.4 FLOW NETWORKS AND LINEAR PROGRAMMING

Networks of flow lend themselves in a natural manner to LP formulations, and conversely: several well-known and time honored LP models lend themselves to representation by flow-networks, and to solution by flow methods.

For instance, consider the flow maximization problem discussed in 2.2 above. Let v denote the total flow out of s (and into t). Let B(j) denote the nodes 'before j' which connect to j, and A(j) denote the nodes 'after j' which connect with j. Then, obviously, maximizing the flow is equivalent to the LP:

$$\text{maximize } v \tag{2-9}$$

$$\text{subject to} \quad \sum_j f(i,j) - \sum_j f(j,i) = \begin{cases} v & \text{if } i=s \\ 0 & \text{if } i\neq s,t \\ -v & \text{if } i=t \end{cases} \tag{2-10}$$

$$0 \leq f(i,j) \leq c(i,j) \quad ,(i,j)\varepsilon A. \tag{2-11}$$

The constraints of (2-10) are the 'flow conservation' constraints, while (2-11) represent the capacity constraints. As we have just witnessed, it would have been foolish to resort to general linear programming theory to find the maximum flow, v*. The simple 'labeling procedure' gave us the desired optimum in an elementary way.

Such simplicity, which, in a real mathematical sense, is synonymous with elegance and beauty, seems to be a characteristic of flow procedures. It seems that a flow interpretation reduces the problem to its bare essentials, which helps bring out the fundamental concepts necessary for its solution.

Of course, not all flow network representations are of an elementary character. Most of them are complicated and require a delicate argument. Most prominent among these is the so-called "general minimal cost circulation problem" which we discuss next in greater detail. As we shall see later on, many practical problems such as transportation, transhipment, assignment, production, project cost, etc. can be recast, i.e., reformulated, as special cases of the minimal cost circulation problem.

2.5 THE GENERAL MINIMAL COST CIRCULATION PROBLEM

Assume given a capacitated network $G = [N,A]$ of $N = (1,2,\ldots,n)$ nodes and A arcs $\{(i,j)\}$. There may be several sources $(s_1,s_2,\ldots,s_m)\equiv S$ with availabilities $a(s_i) > 0$ at source s_i, and several terminals $(t_1,t_2,\ldots t_r)\equiv T$ with demands $a(t_j) < 0$ at terminal t_j. Let $c(i,j)$ denote, as always, the capacity of arc (i,j) in the direction $i \rightarrow j$, and assume the cost to be $b(i,j)$ per unit of the commodity flowing in (i,j) in the direction $i \rightarrow j$. Notice that this implies a linear cost of 'slope' $b(i,j)$. We shall further assume that there exists a lower bound $\ell(i,j) \geq 0$ for each arc (i,j), in the direction $i \rightarrow j$, below which no flow $f(i,j)$ may fall, $0 \leq \ell(i,j) \leq c(i,j)$. In other words, flow in arc (i,j) is bound from below by $\ell(i,j)$ and from above by $c(i,j)$. (This is the most general setup. If no lower bound exists, $\ell(i,j)$ is put $= 0$. If no upper limit on the flow exists, $c(i,j)$ is put $= \infty$.)

It is desired to determine the flow from the sources to the terminals which minimizes the total cost and is within the lower and upper bounds of each arc in the network.[†]

Notice that because of the lower bounds $\ell(i,j)$ together with the capacities $c(i,j)$ there is a question of feasibility, a problem which never arose in maximizing the flow from s to t, discussed in §2.2 above. Therefore, our first task is to address ourselves to this problem: is there a feasible flow in the network?

Let $f(i,N)$ denote the flow out of node i (to the rest of the network) and $f(N,i)$ denote the flow into i (from the rest of the network). Evidently, both flows are through arcs connected to i.

Instead of talking about flow networks we shall talk about circulation networks, i.e., networks which are source and sink free. It is easy to see that any flow network can be transformed into a circulation network by adding enough arcs of the requisite capacities which impose no added restrictions on the original flow network. For instance, the flow network of Fig. 2.9a has two

† Several 'primal' approaches to this problem have been suggested. For a recent contribution and references to others, see ref. [11].

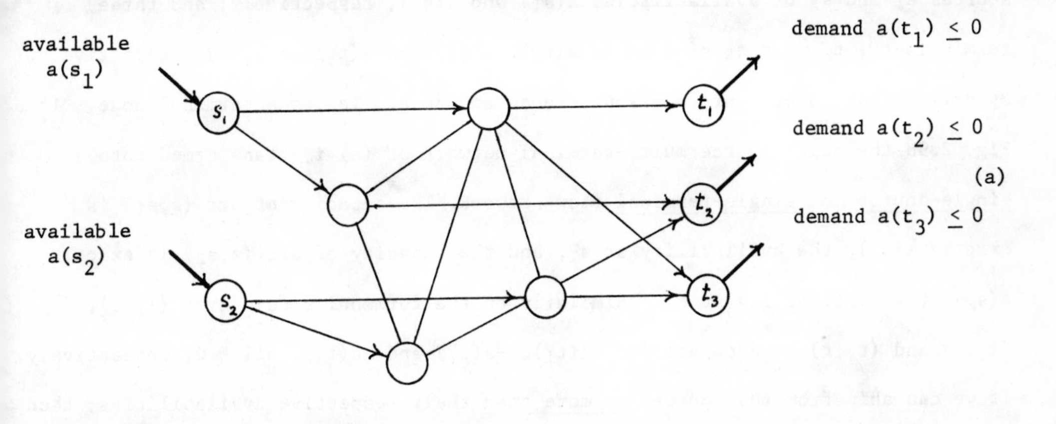
available
a(s1)
available
a(s2)
demand a(t1) ≤ 0
demand a(t2) ≤ 0
demand a(t3) ≤ 0
(a)

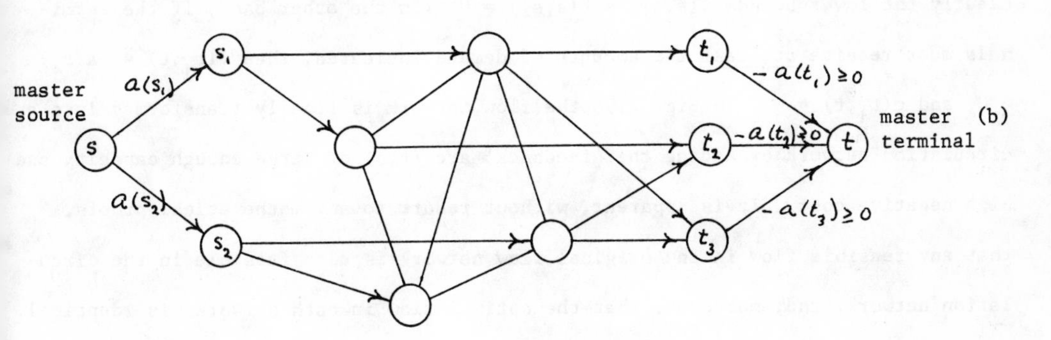
master
source
a(s1)
a(s2)
-a(t1) ≥ 0
-a(t2) ≥ 0
-a(t3) ≥ 0
master
terminal
(b)

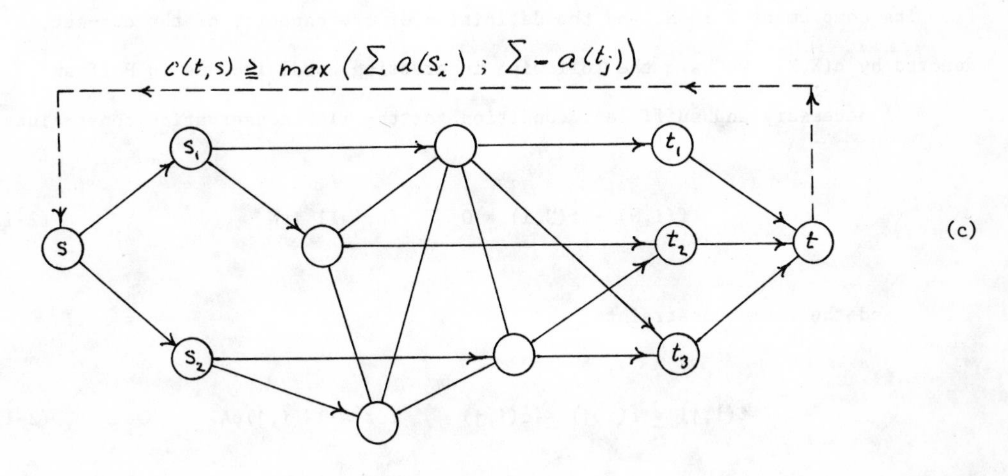
c(t,s) ≥ max (∑ a(si) ; ∑ -a(tj))
(c)

Figure 2.9

sources s_1 and s_2 of availabilities $a(s_1)$ and $a(s_2)$, respectively; and three terminals t_1, t_2, and t_3 of demands $a(t_1)$, $a(t_2)$ and $a(t_3)$, all ≤ 0, respectively. By definition, no arc enters a source node and no arc leaves a terminal node. In Fig. 2.9b the multi-source multi-terminal network of (a) is transformed into a single-source and single-terminal flow network. The capacity of arc (s,s_1) is exactly $a(s_1)$, the availability at s_1, and the capacity of arc (s,s_2) is exactly $a(s_2)$, the availability at s_2. Similarly on the terminal side: arcs (t_1,t), (t_2,t) and (t_3,t) have capacities $-a(t_1)$, $-a(t_2)$ and $-a(t_3)$, all ≥ 0, respectively. If we can ship from the sources no more than their respective availabilities, then clearly the lower bounds $\ell(s,s_1) = \ell(s,s_2) = 0$. On the other hand, if the terminals must receive at least the amounts of demand indicated, then $\ell(t_j,t) = -a(t_j) \geq 0$, and $c(t_j,t) = \infty$. In Fig. 2.9c the flow network is finally transformed into a circulation network by adding the 'feedback' arc (t,s) of large enough capacity and high negative cost. It is apparent, without resort to any mathematical proofs, that any feasible flow in the original flow network is also feasible in the circulation network, and, moreover, that the optimal flow in both networks is identical.

Recalling the definition of a 'cut-set' separating the subset of nodes X from its complement $\bar{X}$ in N, and the definition of the capacity of the cut-set, denoted by $c(X,\bar{X})$, we have the following interesting condition due to Hoffman [9]:

A necessary and sufficient condition for the flow conservation constraints

$$f(i,N) - f(N,i) = 0 \qquad \text{for all } i \in N \tag{2-12}$$

and the bound constraints

$$\ell(i,j) \leq f(i,j) \leq c(i,j) \qquad \text{for all } (i,j) \in A \tag{2-13}$$

to be feasible, where $0 \leq \ell \leq c$, is that

$$c(X,\bar{X}) \geq \ell(\bar{X},X) \tag{2-14}$$

hold for all $X \subseteq N$.

The necessity of the condition is intuitively clear (and easily proved directly), since (2-14) simply asserts that there must be sufficient "escape capacity" from the set X to take care of the flow forced into X by the lower bound ℓ. For otherwise, the conservation of flow equations (2-12) would be violated, since then $\sum_{i\varepsilon X} f(i,N) \leq \sum_{i\varepsilon X} c(i,N) = c(X,\bar{X}) < \ell(\bar{X},X) \leq f(\bar{X},X) = \sum_{i\varepsilon X} f(N,i)$; hence $\sum_{i\varepsilon X} f(i,n) - \sum_{i\varepsilon X} f(N,i) < 0$, contradicting Eqs. (2-12).

The proof of sufficiency is slightly more involved. Extend the network [N,A] to [N',A'] by the adjunction of a source node <u>u</u>, a terminal node <u>w</u>, and a set of arcs (u,N) and (N,w). The lower bound for each added arc is equal to 0, and the capacity for the arcs in A' is given by:

$$\begin{aligned} c'(i,j) &= c(i,j) - \ell(i,j) && (i,j)\varepsilon A \\ c'(u,i) &= \ell(N,i) && i\varepsilon N \\ c'(i,w) &= \ell(i,N) && i\varepsilon N \end{aligned} \tag{2-15}$$

It is easy to verify that corresponding to a <u>feasible circulation</u> f in the original network [N,A] there is a flow f' in the extended network [N',A'] given by:

$$\begin{aligned} f'(i,j) &= f(i,j) - \ell(i,j) \text{ for all } (i,j)\varepsilon A \\ f'(u,i) &= \ell(N,i) && i\varepsilon N \\ f'(i,w) &= \ell(i,N) && i\varepsilon N \end{aligned} \tag{2-16}$$

Figure 2.10 illustrates this fact. Notice that $\sum_{i\varepsilon N} f'(u,i) = \sum_{i\varepsilon N} f'(i,w) = \ell(N,N)$, a constant. This constant flow in [N',A'] is induced between <u>u</u> and <u>w</u> <u>whenever there is a feasible circulation in</u> [N,A]. Thus the question becomes: when is there a flow from <u>u</u> to <u>w</u> in [N',A'] that has the value $\ell[N,N]$? The answer is obtained by invoking the maximum-flow minimum-cut theorem. In particular, a necessary and sufficient condition for the existence of such flow from <u>u</u> to <u>w</u> is that all cut capacities exceed $\ell(N,N)$. Let $(X',\bar{X}')$ be a cut separating <u>u</u> and <u>w</u> in [N',A'], and define $X \subseteq N$ and its complement $\bar{X} \subseteq N$ by

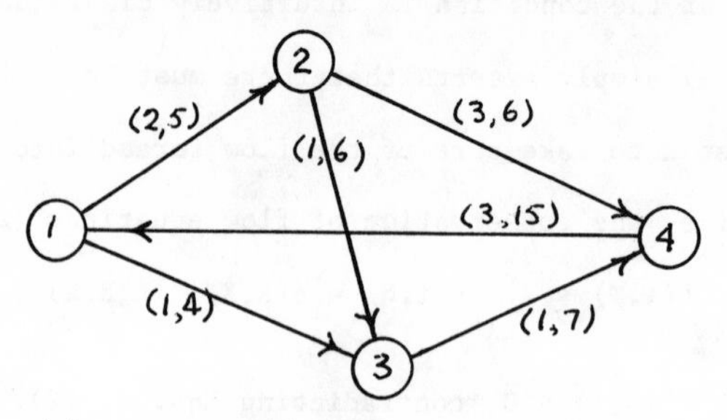

(a) capacitated network; on each arc is defined (ℓ.c.)

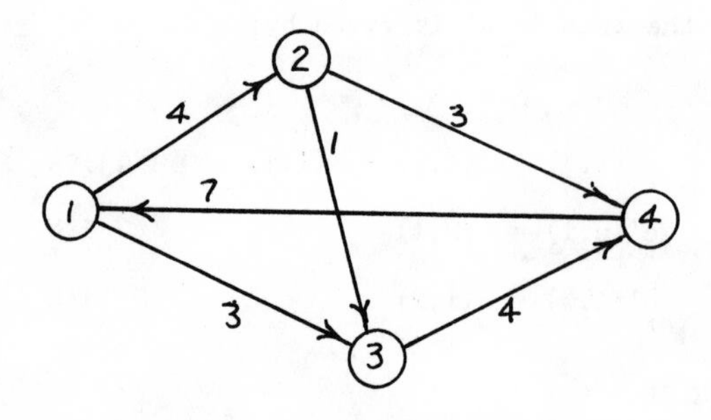

(b) A feasible flow f(i,j) in [N,A]

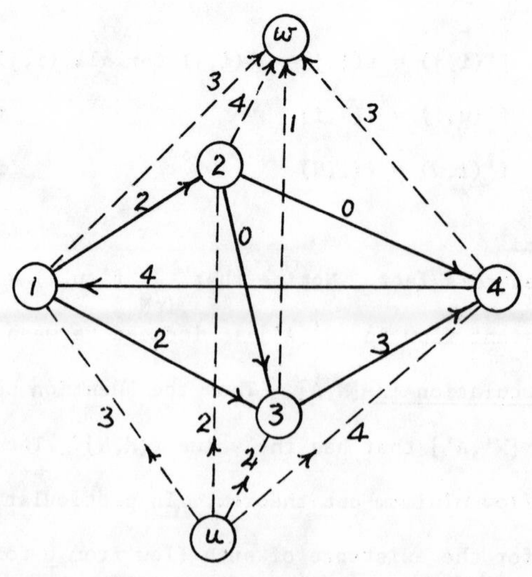

(c) Induced flow from $\underline{u}$ to $\underline{w}$ in [N',A']

Figure 2.10

$$X = X' - u, \ \bar{X} = \bar{X}' - w.$$

Then,

$$\begin{aligned} c'(X',\bar{X}') &= c'(X \cup u, \bar{X} \cup w) \\ &= c'(X,\bar{X}) + c'(u,\bar{X}) + c'(X,w) \\ &= c(X,\bar{X}) - \ell(X,\bar{X}) + \ell(N,\bar{X}) + \ell(X,N) \\ &= c(X,\bar{X}) + \ell(\bar{X},\bar{X}) + \ell(X,N) \end{aligned}$$

which is $\geq \ell(N,N) = \ell(X,N) + \ell(\bar{X},N)$ if and only if

$$c(X,\bar{X}) \geq \ell(\bar{X},X)$$

which is condition (2-14).

It is important to remark that the necessary and sufficient condition of (2-14) is valid only for directed networks but not for undirected or mixed networks. In general, replacing an undirected arc by a pair of oppositely directed arcs and cancelling flows in opposite directions is not a valid operation.

Of course, this condition is useful only as a check rather than as a constructive method to verify feasibility. In other words, it is not computationally feasible to enumerate all the cut-sets in [N,A] and check that (2-14) is satisfied. Rather, we start the solution of our minimum cost circulation problem assuming that flow is feasible. If, in fact, the problem is infeasible, then, at some stage of the iteration, condition (2-14) will be violated and calculation stops.

We return now to our original cost minimization problem, which can be succinctly stated as:

$$\text{minimize} \sum_{A} b(i,j)\ f(i,j) \tag{2-17}$$

subject to constraints

$$f(i,N) - f(N,i) = 0 \qquad \text{all } i\varepsilon N \qquad (2\text{-}18)$$

$$0 \leq \ell(i,j) \leq f(i,j) \leq c(i,j) \qquad \text{all } (i,j)\varepsilon A \qquad (2\text{-}19)$$

A fundamental observation, whose understanding is crucial for grasping the intricate workings of this algorithm as well as almost all other algorithms related to flow network problems, is the well-known complementary slackness theorem (CST) of linear programming. This theorem simply states that:

> To each primal linear program defined by: minimize $z = cX$ subject to $AX \geq b$, $X \geq 0$, where c, X and b are vectors and A is an m×n matrix, there corresponds a dual linear program defined by: maximize $g = bY$ subject to $A^tY \leq c$, $Y \geq 0$. At the optimal, if it exists, if a dual variable y_j is > 0 then it must be true that the corresponding jth inequality in the primal is satisfied as equality. Since the primal LP is the dual of the dual LP, the theorem also implies that if some x_j is > 0 then the corresponding jth dual constraint must be satisfied as equality.

We now apply this theorem to the LP of (2-17) - (2-19). But first, let us write the LP in the following fashion, in which we also exhibit the dual variables:

$$\text{minimize } z = \sum_A b(i,j)\ f(i,j)$$

subject to

		Dual Variables
$-f(i,N) + f(N,i) = 0$	, all $i\varepsilon N$	π_i
$f(i,j) \geq \ell(i,j)$	, all $(i,j)\varepsilon A$	δ_{ij}
$-f(i,j) \geq -c(i,j)$	, all $(i,j)\varepsilon A$	γ_{ij}

Formulating the dual LP we immediately obtain the typical dual constraint

$$-\pi_i + \pi_j + \delta_{ij} - \gamma_{ij} \leq b(i,j), \quad \text{all } (i,j)\varepsilon A \qquad (2\text{-}20)$$

By the CST, if $\delta_{ij} > 0 \Rightarrow f(i,j) = \ell(i,j)$

if $\gamma_{ij} > 0 \Rightarrow f(i,j) = c(i,j)$ (2-21)

if $f(i,j) > 0 \Rightarrow -\pi_i + \pi_j + \delta_{ij} - \gamma_{ij} = b(i,j)$, all $(i,j)\varepsilon A$

Consequently, at the optimal (if it exists), it is impossible for both δ_{ij} and γ_{ij} to be > 0[†], though <u>both may be zero</u>, and in that case we must have $\ell(i,j) \leq f(i,j) \leq c(i,j)$. Moreover, if we let $\bar{b}(i,j) = b(i,j) + \pi_i - \pi_j \geq \delta_{ij} - \gamma_{ij}$, then clearly

$$\text{if } \bar{b}(i,j) > 0 \Rightarrow \delta_{ij} > 0 \Rightarrow f(i,j) = \ell(i,j)$$

and (2-22)

$$\text{if } \bar{b}(i,j) < 0 \Rightarrow \gamma_{ij} > 0 \Rightarrow f(i,j) = c(i,j)$$

These are the optimality conditions, which are both necessary and sufficient, towards which the algorithm to be described works. In other words, we are seeking flows $f(i,j)$ and node numbers (π_i) such that the conditions (2-22) are satisfied. The dual variables δ and γ do not appear in an explicit fashion.

To sum up the above discussion, if we have positive flows f and node numbers π such that each arc (i,j) with $f(i,j) > 0$ is in one of three states:

(1) $\bar{b}(i,j) > 0,\ f(i,j) = \ell(i,j)$

(2) $\bar{b}(i,j) = 0,\ \ell(i,j) \leq f(i,j) \leq c(i,j)$ (2-23)

(3) $\bar{b}(i,j) < 0,\ f(i,j) = c(i,j)$

then we know we have an <u>optimal</u> flow. The optimality conditions can be violated in several ways, which we enumerate:

$$\bar{b}(i,j) > 0 \text{ and } f(i,j) < \ell(i,j) \quad (2\text{-}24)$$

$$\text{or } f(i,j) > \ell(i,j) \quad (2\text{-}25)$$

$$\bar{b}(i,j) = 0 \text{ and } f(i,j) < \ell(i,j) \quad (2\text{-}26)$$

$$\text{or } f(i,j) > c(i,j) \quad (2\text{-}27)$$

$$\bar{b}(i,j) < 0 \text{ and } f(i,j) < c(i,j) \quad (2\text{-}28)$$

$$\text{or } f(i,j) > c(i,j) \quad (2\text{-}29)$$

---

† Except for the special case in which $\ell = c$, and then the flow f is specified <u>a priori</u>.

The alert reader will notice that conditions (2-24), (2-26), (2-27) and (2-29) denote infeasible solutions. Their inclusion in the algorithm renders it more general, since one can start with any flow, feasible or infeasible, and any node numbers Π. Moreover, if it is impossible to remove the infeasibility, then the problem possesses no solution. Thus the algorithm provides the operational test for infeasibility referred to in our discussion of condition (2-14). As we shall see below, it is preferable to start the iterations with a feasible flow, if at all possible, since this would eliminate four out of the six non-optimal conditions throughout the calculations, with concomitant savings in time and effort.

The algorithm we are about to describe is called the 'out-of-kilter' method and is due to Fulkerson[6]†. It has the interesting property that the status of no arc of the network is worsened at any step of the computation. Thus, if one starts with a feasible flow in any arc, the flow will remain feasible in that arc. In fact, if one starts with an optimal flow in any arc (i.e., the flow and node numbers satisfy the optimality conditions of (2-23)) the flow will remain optimal thereafter in that arc. Hence, it is better to start as close to the optimal as possible; for example, with a feasible flow. The objective of the analysis, of course, is to change the arc flows and node numbers to satisfy the optimality conditions of (2-23).

With any flow vector $\underline{f}$ and set of node numbers Π, an arc either satisfies the optimality conditions (2-23), and is said to be in kilter, or it does not satisfy these conditions, and is said to be out of kilter.

As always, we assume that ℓ, c and b are all integers. The procedure is initiated with any integer circulation $\underline{f}$ and a set of integer node numbers Π. The algorithm is depicted in Fig. 2.11, and its application is illustrated in Fig. 2.12 and the associated Table 2.1.

---

† The title of the procedure was coined by Fulkerson. According to the Concise Oxford dictionary, 'out-of-kilter' is English dialect meaning "not working properly." The opposite is 'in-kilter' meaning "in good working order."

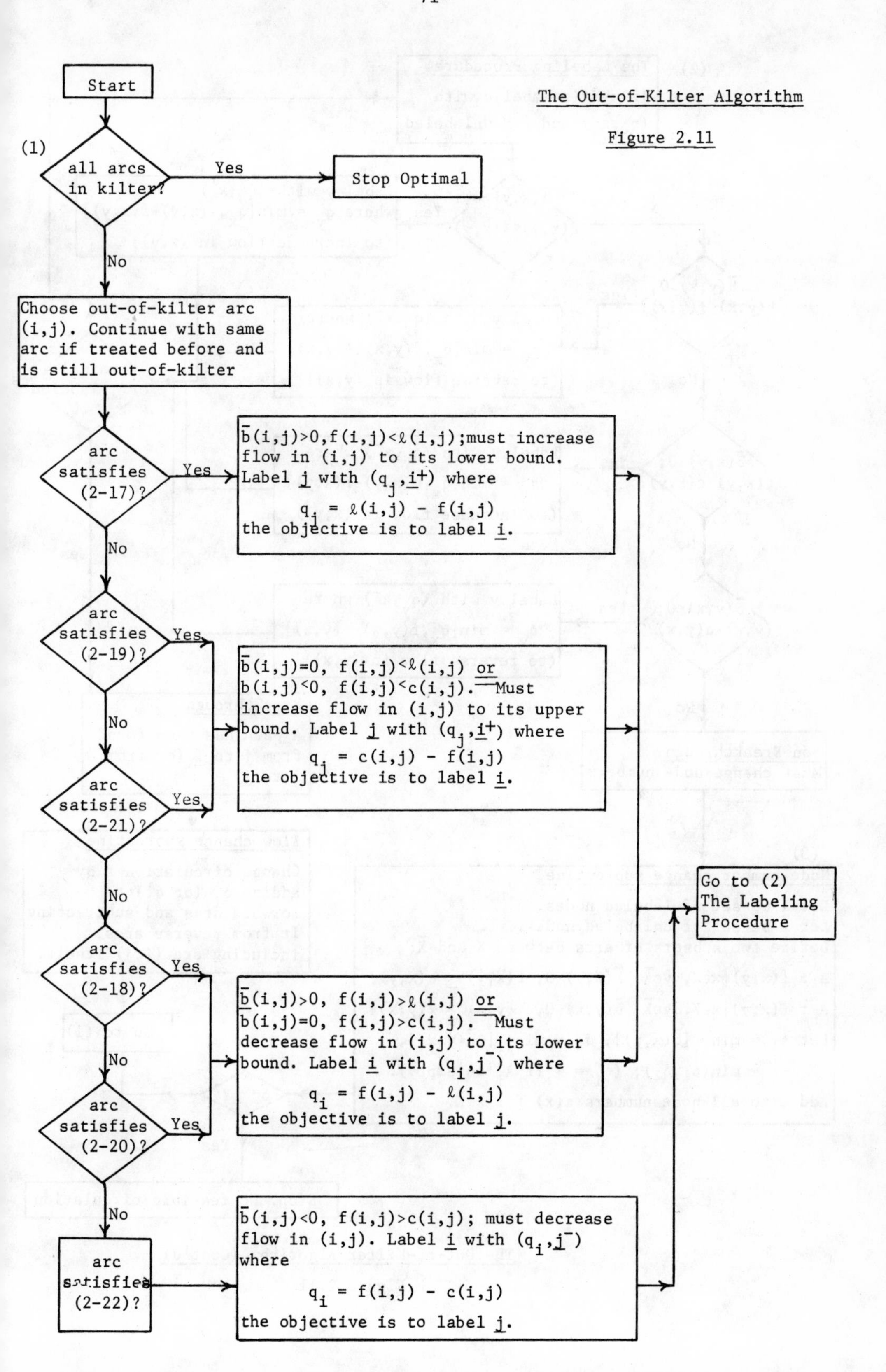

The Out-of-Kilter Algorithm

Figure 2.11

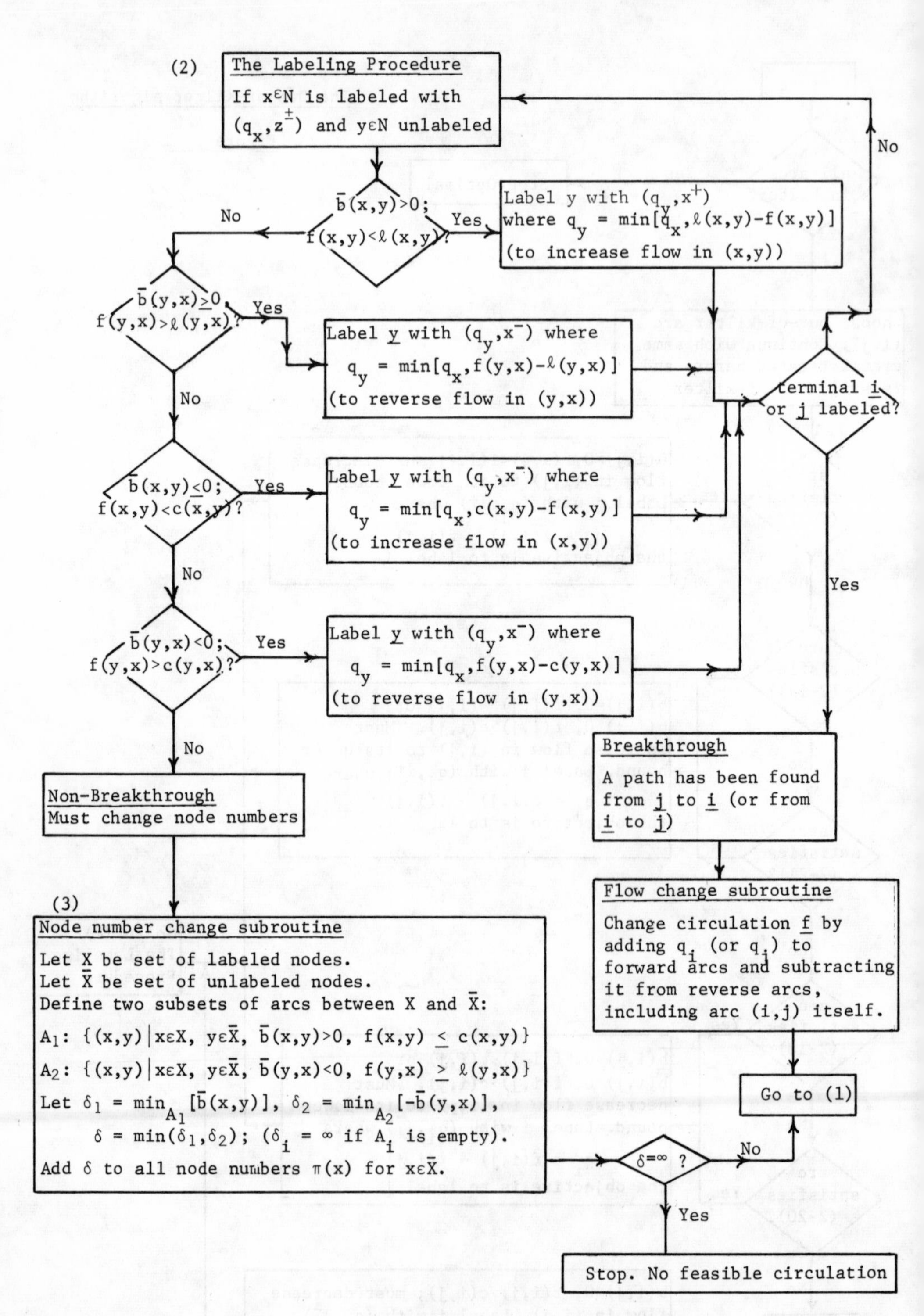

The Out-of-Kilter Algorithm (Cont'd)

Figure 2.11

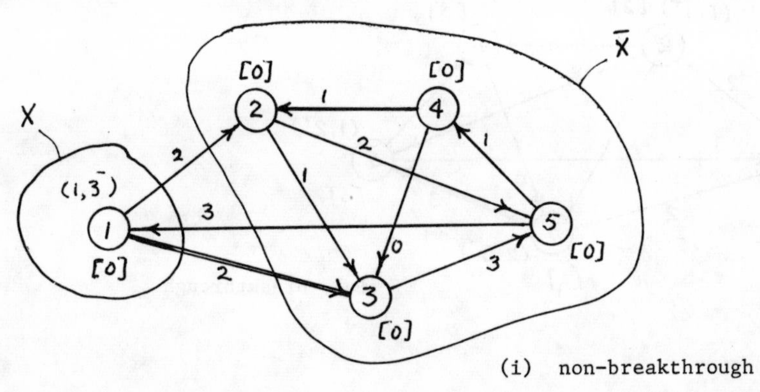

(i) non-breakthrough

$A_1 = \{(1,2),(1,3)\}, A_2 = \{(5,1)\}, \delta = 1$

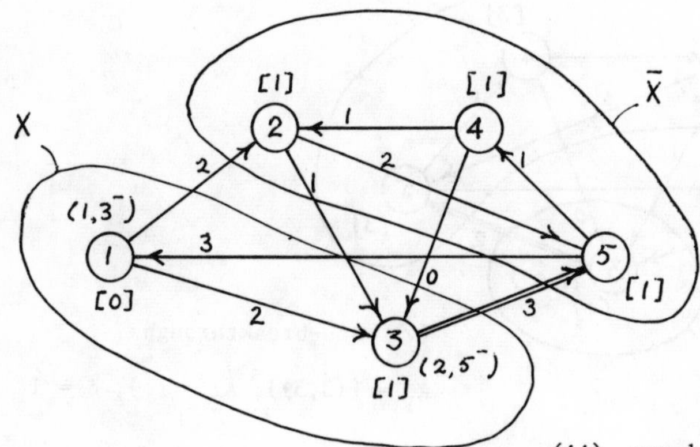

(ii) non-breakthrough

$A_1 = \{(1,2),(3,5)\}, A_2 = \{(5,1)\}, \delta = 1$

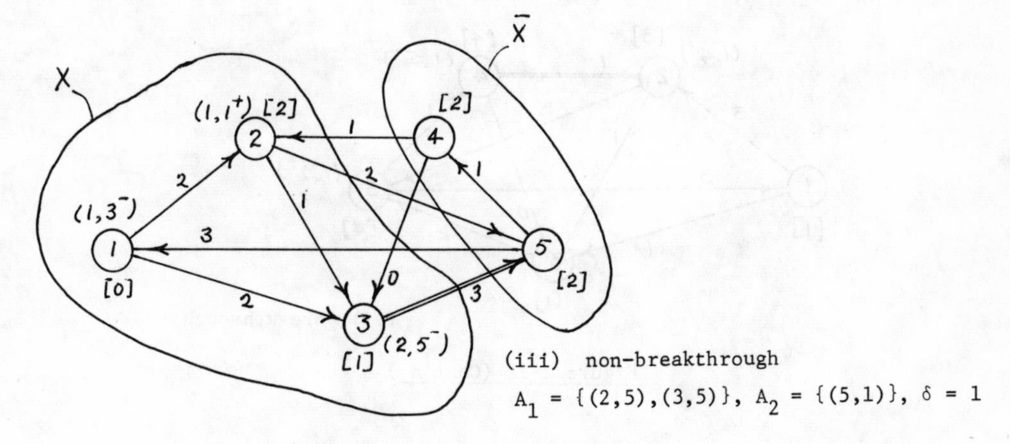

(iii) non-breakthrough

$A_1 = \{(2,5),(3,5)\}, A_2 = \{(5,1)\}, \delta = 1$

Figure 2-12 Example of Out-of-Kilter Method

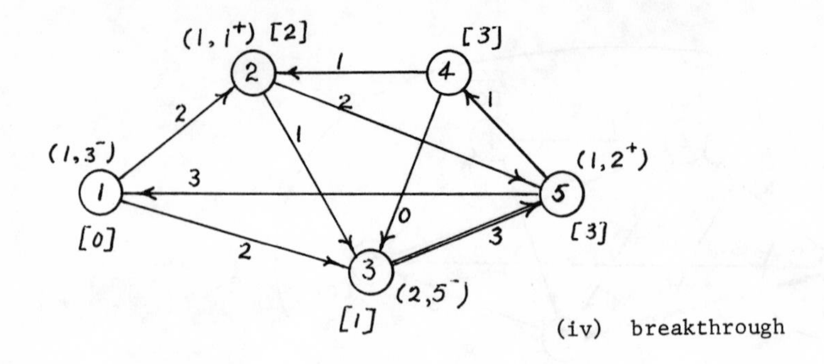

(iv) breakthrough

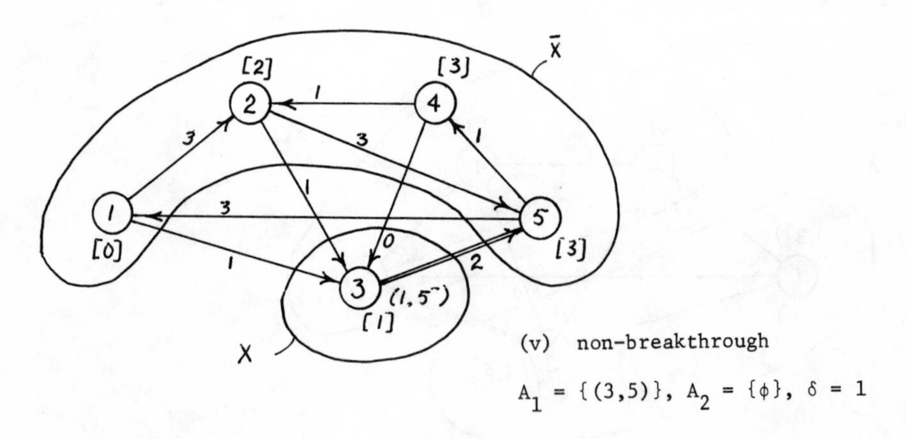

(v) non-breakthrough

$A_1 = \{(3,5)\}$, $A_2 = \{\phi\}$, $\delta = 1$

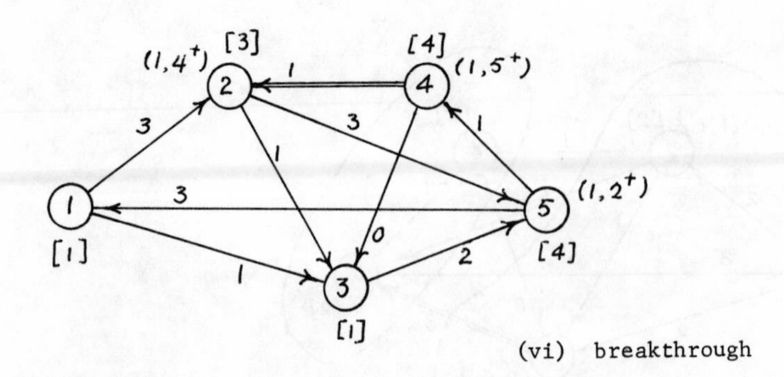

(vi) breakthrough

Figure 2-12 (Cont'd.)

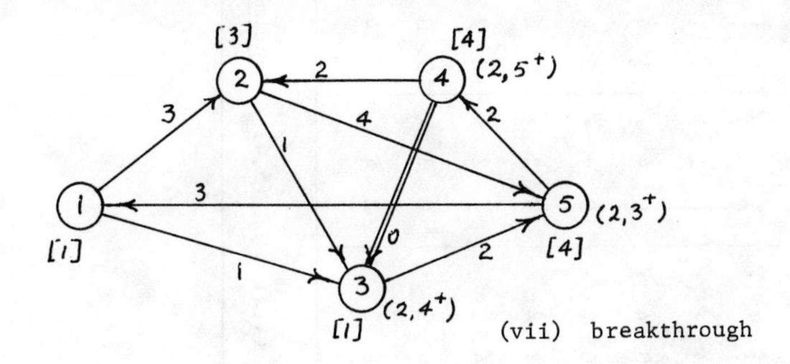

(vii) breakthrough

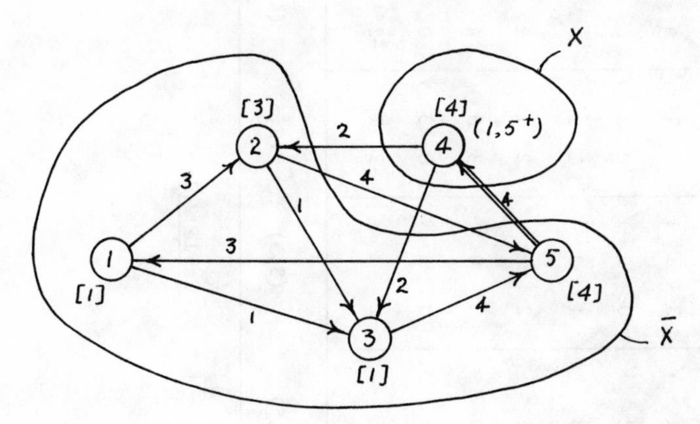

(viii) non-breakthrough

$A_1 = \{(4,2),(4,3)\}$, $A_2 = \{(5,4)\}$, $\delta = 2$

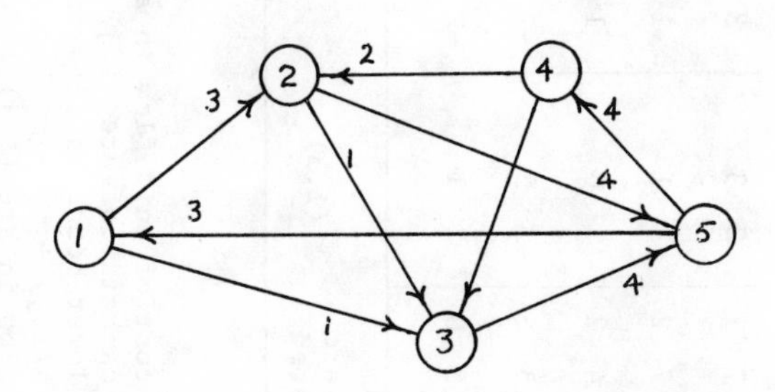

(ix) Optimal Circulation: cost = 12

<u>Example of Out-of-Kilter Method</u>

<u>Figure 2-12 (Cont'd)</u>

	Basic data			(i)			(ii)			(iii)			(iv)			(v)			(vi)			(vii)			(viii)			(ix)		
arc (i,j)	b	c	ℓ	$\bar{b}$	f	k	$\bar{b}$	f	k	$\bar{b}$	f	k	$\bar{b}$	f	k	$\bar{b}$	f	k	$\bar{b}$	f	k	$\bar{b}$	f	k	$\bar{b}$	f	k	$\bar{b}$	f	k
(1,2)	2	5	2	2	2	√	1	2	√	0	2	√	0	2	√	0	3	√	0	3	√	0	3	√	0	3	√	0	3	√
(1,3)	1	4	1	1	2		0	2	√	0	2	√	0	2	√	0	1	√	1	1	√	1	1	√	1	1	√	1	1	√
(2,3)	2	6	1	2	1	√	2	1	√	3	1	√	3	1	√	3	1	√	4	1	√	4	1	√	4	1	√	4	1	√
(2,5)	1	6	2	1	2	√	1	2	√	1	2	√	0	2	√	0	3	√	0	3	√	0	4	√	0	4	√	0	4	√
(3,5)	3	7	1	3	3		3	3		2	3		1	3		1	2		0	2	√	0	2	√	0	4	√	0	4	√
(4,2)	1	7	2	1	1		1	1		1	1		2	1		2	1		2	1		2	2	√	2	2	√	0	2	√
(4,3)	1	4	2	1	0		1	0		2	0		3	0		3	0		4	0		4	0		4	2	√	2	2	√
(5,1)	-3	15	3	-3	3		-2	3		-1	3		0	3	√	0	3	√	0	3	√	0	3	√	0	3	√	0	3	√
(5,4)	-2	5	0	-2	1		-2	1		-2	1		-2	1		-2	1		-2	1		-2	2		-2	4		0	4	√
out-of-kilter arc being investigated				(1,3)			(3,5)			(3,5)			(3,5)			(3,5)			(4,2)			(4,3)			(5,4)			optimal		

Table 2.1

Symbols:
- b: cost of unit flow in arc (i,j)
- c: capacity of arc (i,j)
- ℓ: lower bound on flow in arc (i,j)
- $\bar{b}$= $a(i,j) + \pi_i - \pi_j$
- f: flow in arc (i,j)
- k: in-kilter arcs

The formal proof of the procedure is rather delicate and involved, especially the proof of its finiteness. The interested reader may consult the article of Fulkerson[6] or pages 162-169 of the book by Ford and Fulkerson[5]. However, the heuristic justification for each step is rather simple and intuitively appealing. The reader must remember always that the objective is to satisfy the complementary slackness optimality conditions of (2-23). These, or similar conditions, are the cornerstone of almost all flow algorithms. Consider, for example, an out-of-kilter arc, say arc (3,5) in iteration (iv) of Fig. 2-12: $\bar{b}(3,5) > 0$ but the flow $f(3,5) = 3 > \ell(3,5) = 1$. Since $\bar{b}$ represents the imputed net cost of the arc, $\bar{b} > 0$ implies a net imputed loss. Consequently, it is better to <u>reduce</u> the flow in such an arc to its lowest possible level, i.e., to $\ell(3,5) = 1$. But reduction of flow can be achieved only through the <u>reversal</u> of the excess (or unwanted) flow in that arc, hence the labeling of $\underline{i}$ with (q_i, j^-) where $q_i = f(i,j) - \ell(i,j)$. The negative sign on the j is simply a reminder of this reversal of flow.

The conservation of flow obliges us to determine a path from $\underline{i}$ to $\underline{j}$ which would sustain the reversed flow, if such flow is at all feasible. Such feasibility is tested through the labeling procedure. Certainly, if the path contains a <u>forward arc</u> (x,y) (i.e., an arc whose orientation is from $\underline{x}$ to $\underline{y}$), one is allowed to <u>increase</u> the flow in such an arc under two conditions only (and still comply with the optimality conditions of (2-23)):

$$\text{either } \bar{b}(x,y) > 0 \text{ and } f(x,y) < \ell(x,y)$$
$$\text{or } \bar{b}(x,y) \leq 0 \text{ and } f(x,y) < c(x,y)$$

In the first case, the flow may be increased up to $\ell(x,y)$ and no more. In the second case, the flow may be increased up to $c(x,y)$ and no more. On the other hand, if the path contains a <u>reverse arc</u> (x,y) (i.e., an arc whose orientation is from $\underline{y}$ to $\underline{x}$), then one must be <u>reversing</u> some flow in that arc. Again, by the conditions (2-23), such reversal is permitted only if:

$$\text{either } \bar{b}(y,x) < 0 \text{ and } f(y,x) > c(y,x)$$
$$\text{or } \bar{b}(y,x) \geq 0 \text{ and } f(y,x) > \ell(y,x)$$

In the first case, the flow may be reduced to $c(y,x)$ but not any lower, and in the second case it may be reduced to $\ell(y,x)$ but not any lower.

If the reader would compare the last two paragraphs with the chart of the labeling procedure, Fig. 2.11, he will find them to be identical. The labeling with the minimum of either availability at the source node or the permissible flow increase (or reversal) is simply to guarantee that the status of no arc is worsened (i.e., it is removed from in-kilter to out-of-kilter condition). Now the labeling of Fig. 2.12 (iv) should be obvious. Since breakthrough occurred, a closed circuit must have been found from node 3 back to node 3 *via* node 5. The labeling indicates that:

> flow, of a quantity $q_5 = 1$, in arcs (3,5) and (1,3) must be reversed; and flow, of the same quantity, in arcs (1,2) and (2,5) must be increased.

This was done and provided the starting point of iteration (v).

A similar intuitively appealing argument can be presented for changing the node numbers. Here, it is sufficient to remark that an arc (x,y) in the cut-set $C(X,\bar{X})$, $x \varepsilon X$ and $y \varepsilon \bar{X}$, (i.e., x labeled and y not labeled) is either oriented from X into $\bar{X}$ or conversely, from $\bar{X}$ into X:

(1) orientation $x \to y$: it was not possible to label y from x either because $\bar{b}(x,y) > 0$ but $f(x,y) > \ell(x,y)$, or because $\bar{b}(x,y) \leq 0$ but $f(x,y) > c(x,y)$.

In the first case, increasing the node number of y by $\bar{b}(x,y)$ renders the arc in-kilter (because then $\bar{b}(x,y) = 0$ and $c(x,y) \geq f(x,y) > \ell(x,y)$). In the second case, no change in node numbers, by itself, can make the arc in-kilter, hence it does not influence the decision.

(2) orientation $y \rightarrow x$: it was not possible to label $\underline{y}$ from $\underline{x}$ either because $\bar{b}(y,x) \geq 0$ but $f(y,x) \leq \ell(y,x)$ or because

$$\bar{b}(y,x) < 0 \text{ but } f(y,x) < c(y,x).$$

In the first case, no change in node numbers, by itself, can make the arc in-kilter; hence it does not influence the decision. In the second case, increasing the node number of $\underline{y}$ by $\bar{b}(y,x)$ renders the arc in-kilter (because then $\bar{b}(y,x) = 0$ and $\ell(y,x) \leq f(y,x) < c(y,x)$).

We thus reach the conclusion that the node number of $\underline{y}$ should be increased by $\bar{b}(x,y)$ or $\bar{b}(y,x)$ if either $f(x,y) > \ell(x,y)$ or $f(y,x) < c(y,x)$, respectively. The reader can verify that this is precisely the condition in the node number change subroutine. The evaluation of the δ is simply to determine the minimal such increase in node numbers in the set $\bar{X}$ that guarantees no worsening of the status of any arc (i.e., from an in-kilter to out-of-kilter condition).

§ 2.6 SOME APPLICATIONS

The out-of-kilter method is thus seen to be a potent and computationally tractable procedure for determining the least cost flow in capacitated networks, when cost is a linear function of the flow. We now proceed with the translation of a few well-known non-flow problems mentioned in the preamble to this chapter into flow format, to which the above procedure, or simple variations thereof, is applicable.

1. The Personnel Assignment Problem

Represent each graduate by a 'source' node $\underline{s}_i$, $i = 1,\ldots,n$, and connect a master source $\underline{s}$ to $\underline{s}_i$, $i = 1,\ldots,n$, with a directed arc of capacity 1. Represent each vacant job by a 'terminal' node $\underline{t}_j$, $j = 1,\ldots,n$, and connect each terminal $\underline{t}_j$ to a master terminal $\underline{t}$ with a directed arc of capacity 1. Now connect each source $\underline{s}_i$ to each terminal $\underline{t}_j$ with a directed arc (s_i,t_j) of 'cost'$-a_{ij}$ equal to the negative of the 'proficiency index' of candidate i on job j. The capacity of arc (s_i,t_j) is put equal to ∞. The network for n = 4 is shown in Fig. 2.13.

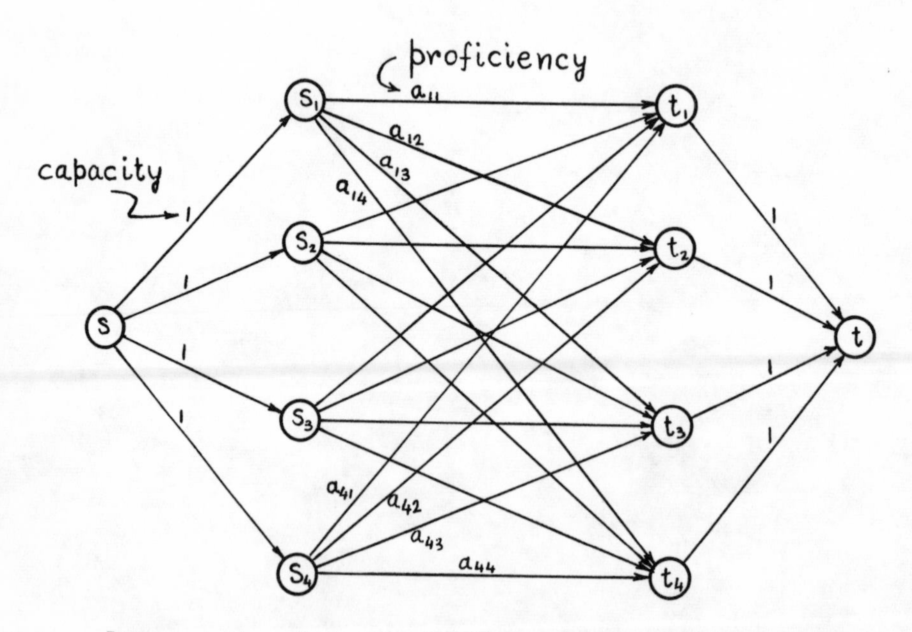

Representation of the Personnel Assignment Problem in Flow-Network

Figure 2.13

This is a cost-of-flow minimization problem in a capacitated network for which the lower bound $\ell(i,j) = 0$ for all the arcs of the network. It can be solved by the out-of-kilter method.

2. A Production-Inventory Problem

Assume both the cost of production $p_t(x_t)$ and the inventory holding cost $h_t(I_t)$ to be both linear. The flow-network representation for the case of 4 time periods is shown in Fig. 2-14. Node p is a source of production, and each arc (p,p_t) has lower bound 0, and upper bound c_t equal to the productive capacity in period t, and a cost per unit produced of a_t. The inventory at the end of period t is I_t, and is represented by the flow in arc (p_t, p_{t+1}). There is a limit on the quantity stored, γ_t, and the cost of storage is α_t. Finally, sale in period t is flow in arc (p_t,q_t), which has infinite capacity and no cost. The master 'terminal' q connects with each node q_t; the arc (q_t,q) has lower bound b_t, infinite capacity and a unit cost equal to $-s_t$ (i,e., income).

Again, this is a capacitated network in which the value of the flow from p to q can be optimized by the out-of-kilter method.

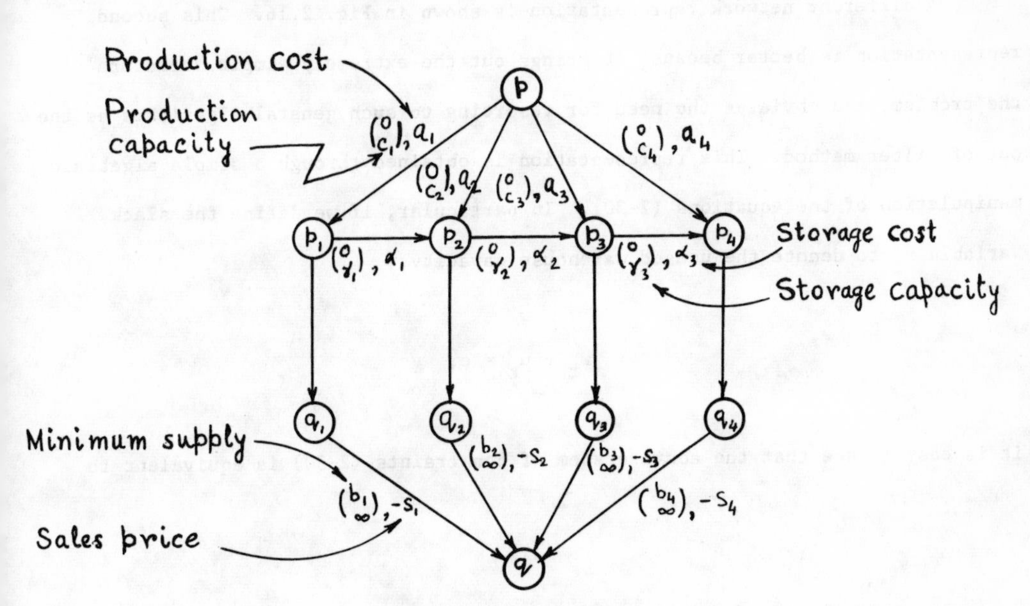

Representation of the Production-Inventory Problem in Flow-Network

Figure 2.14

3. The Warehousing Problem

It is evident that this problem is very similar to the Production-Inventory problem discussed above, with the added specification that the only constraint in the problem is the warehouse capacity.

A possible network representation is shown in Fig. 2.15 for the case T = 3 time periods. This is an obvious representation of the equations:

$$h_{t-1} + q_t = w_t$$
$$h_t = w_t - s_t \qquad (2\text{-}30)$$
$$w_t \leq c$$

where q_t is the quantity purchased, h_{t-1} is the quantity held over from the previous period, s_t is the quantity sold and w_t is the total quantity available at the beginning of the period. The capacity of the warehouse is taken to be c; (the same symbols are used to represent the cost of the activity in Fig. 2-15). All variables are ≥ 0.

A different network representation is shown in Fig. 2.16. This second representation is better because it brings out the extremely simple structure of the problem, and obviates the need for resorting to such general algorithms as the out-of-kilter method. This representation is obtained through a simple algebraic manipulation of the equations (2-30). In particular, if we define the slack variable u_i to denote the unused warehouse capacity,

$$w_t + u_t = c,$$

it is easy to see that the above system of constraints (2-30) is equivalent to

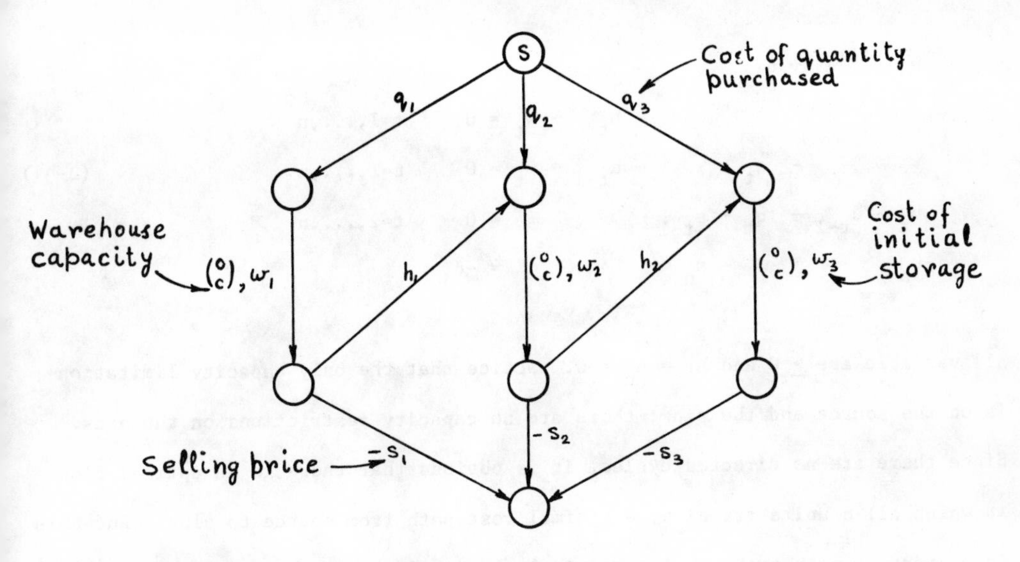

Representation of the Warehousing Problem in Flow-Network

Figure 2.15

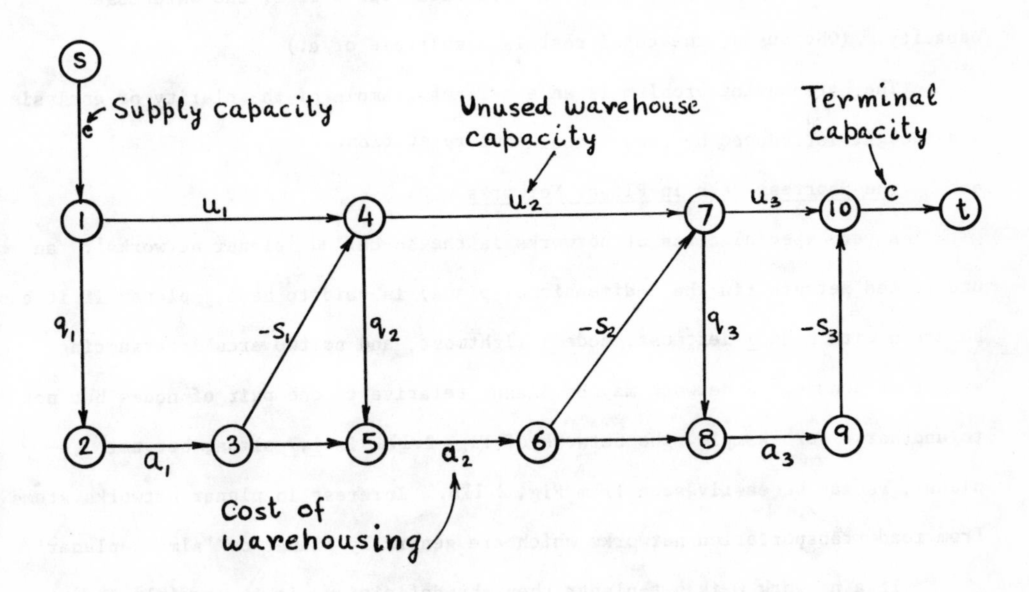

Second Representation of the Warehousing Problem in Flow-Network

Figure 2.16

$$
\begin{array}{llllllll}
u_1 & + q_1 & & & & = c & & \\
& & s_t & + h_t & - w_t & = 0 & t=1,\ldots,n & \\
& - q_t & & - h_{t-1} & + w_t & = 0 & t=1,\ldots,n & (2\text{-}31) \\
u_t - u_{t-1} & + q_t & - s_{t-1} & & & = 0 & t=2,\ldots,n & \\
-u_n & & - s_n & & & = -c, & &
\end{array}
$$

all variable are ≥ 0 and $h_0 = h_n = 0$. Notice that the only capacity limitation is on the source and the sink; there are no capacity restrictions on the arcs. Since there are no directed cycles, it is obvious that there is an optimal flow in which all c units travel by a minimal cost path from source to sink. But this is a shortest path problem of a particularly simple kind, and its solution is readily available. Notice that a single path will be selected. Hence, the optimal pattern of buying and selling is of the 'all or nothing' kind, that is, whatever action is taken in a period is pursued to the limit of warehouse capacity. Such a pattern is really independent of the numerical value of c, the warehouse capacity. (Obviously, the total cost is a multiple of c.)

The Warehousing problem is an excellent example of the clarity of analysis and thought introduced by flow-network interpretation.

4. The Shortest Path in Planar Networks

A very special class of networks is the so-called 'planar networks': an undirected network (in the 2-dimensional plane) is said to be i,j-planar if it can be drawn with node i leftmost, node j rightmost, and no two arcs intersecting except at a node. A network may be planar relative to one pair of nodes but not to another. For example, the network of Fig. 2.17a is $1,7$-planar but not $3,6$-planar, as can be easily seen from Fig. 2.17b. Interest in planar networks stems from road transportation networks which are generally planar or 'almost planar'.

If a network G is s,t-planar then, by definition, it is possible to draw a line extending from node s to the left and a line from node t to the right. These two lines (assumed to extend indefinitely) define two areas: the 'top' area, which

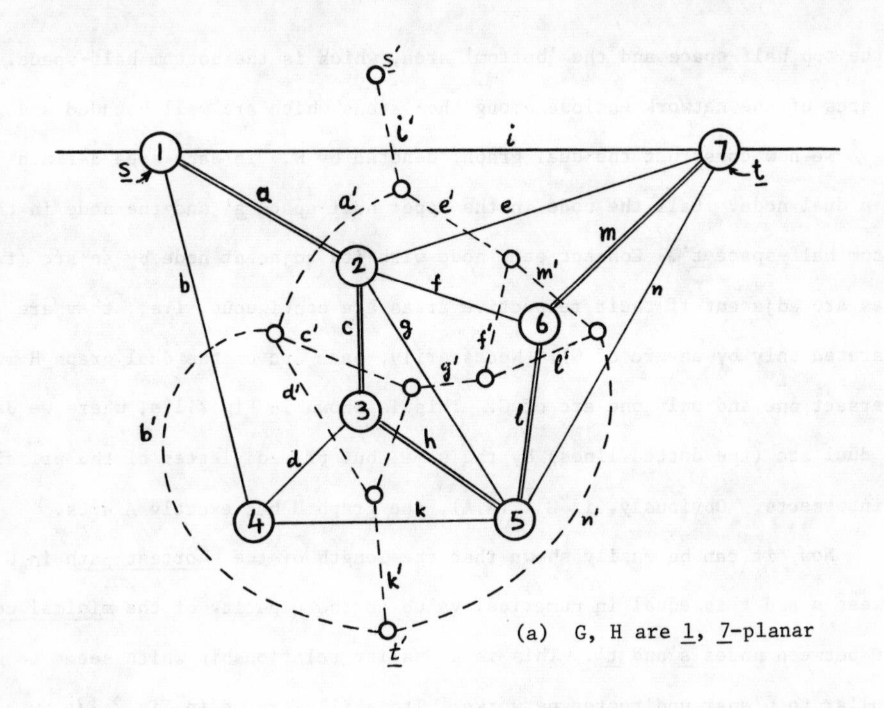

(a) G, H are 1, 7-planar

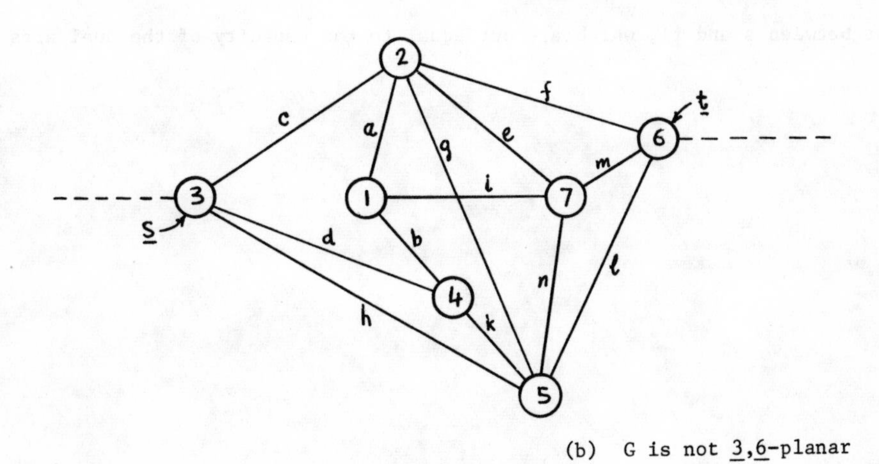

(b) G is not 3,6-planar

Figure 2-17

is the top half-space and the 'bottom' area, which is the bottom half-space. The arcs of the network enclose among them areas which are well bounded and finite

We now construct the dual graph, denoted by H. In each area defined by G put a dual node. Call the node in the upper half-space s' and the node in the bottom half-space t'. Connect each node with its adjacent node by an arc (two nodes are adjacent if their respective areas are contiguous, i.e., they are separated only by an arc of G). Necessarily, each arc of the dual graph H must intersect one and only one arc of G. This is shown in Fig.2.17a, where we denoted the dual arc (the dotted lines) by the same, but primed, letter of the primal arc it intersects. Obviously, if G ≡ (N,A), the graph H has exactly A arcs.

Now, it can be easily shown that the length of the shortest path in G between s and t is equal in numerical value to the capacity of the minimal cut-set in H between nodes s'and t'. This is a duality relationship which seems to be peculiar to planar undirected networks. It is illustrated in Fig.2.17a by assuming that the shortest path between nodes 1 and 7 is composed of arcs a,c,h,ℓ,m. It is easy to see that the dual arcs a',c',h',ℓ',m' constitute a cut-set between s' and t' in H. That the capacity of this cut-set is minimal awaits assigning numerical values to the length of the arcs in G (such that the indicated path is the shortest between s and t), which are put equal to the capacity of the dual arcs in H.

§2.7 CONCAVE COST FUNCTIONS

Loosely speaking, a concave function looks like an inverted bowl. Another way of characterizing it is to say that the chord joining any two points $f(x_1)$ and $f(x_2)$ lies wholly on or below the trace of the function $f(x)$ as x varies between x_1 and x_2, see Fig. 2.18a.

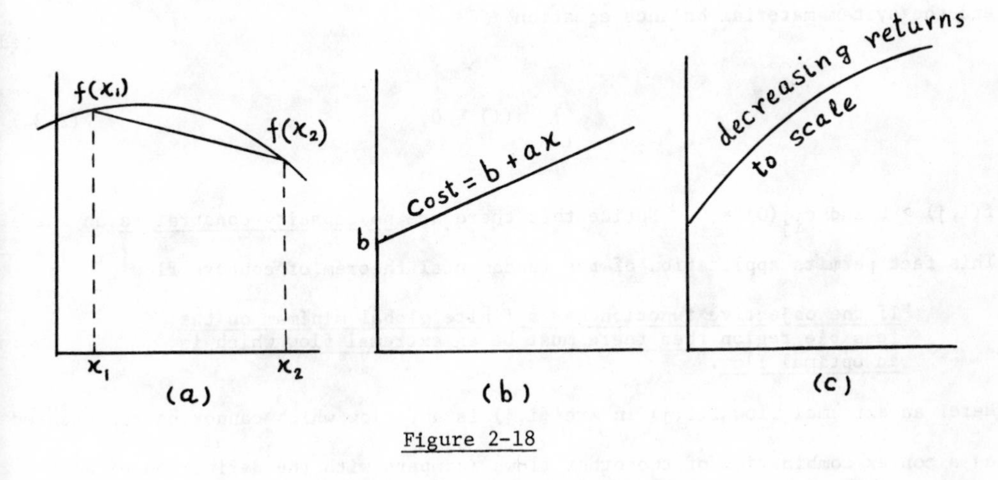

Figure 2-18

In many instances it is more meaningful to speak of concave cost functions than of linear functions. This is true where there is a setup cost (sometimes called the 'fixed charge'), even though the cost may be linear for any positive level of the activity, as shown in Fig. 2.18b. It is also true where there is discounting or efficiencies of scale, as shown in Fig. 2.18c.

Networks of flow with concave cost functions are virtually virgin territory, so little have they been studied. In the following we present some recent results due to Zangwill[12].

The formulation of the problem is as follows. Let $f(i,j)$ denote the flow in arc (i,j) in the direction $\underline{i}$ to $\underline{j}$, $A(i)$ and $B(i)$ the set of nodes 'after $\underline{i}$' and 'before $\underline{i}$', respectively, which connect to $\underline{i}$, $a(i)$ the availability at $\underline{i}$ (a negative availability denotes demand), and $c_{ij}(f)$ a concave cost function of the flow $f(i,j)$; a negative c_{ij} indicates profit or income. Then it is desired to

$$\text{minimize} \sum_{(i,j)\varepsilon A} c_{ij}(f) \tag{2-32}$$

subject to the flow conservation constraints

$$\sum_{j\varepsilon A(i)} f(i,j) - \sum_{j\varepsilon B(i)} f(j,i) = a(i), \quad \text{all } i\varepsilon N \tag{2-33}$$

and the system material balance equation

$$\sum_{i\varepsilon N} a(i) = 0, \tag{2-34}$$

$f(i,j) \geq 0$ and $c_{ij}(0) = 0$. Notice that there are <u>no capacity constraints</u> anywhere. This fact permits application of the fundamental theorem of concave flow [3]:

> "<u>If the objective function has a finite global minimum on the feasible region then there must be an extremal flow which is an optimal flow.</u>"

Here, an extremal flow $f(i,j)$ in arc (i,j) is any flow which cannot be represented as a convex combination of two other flows (compare with the definition of an extreme point in a convex set). In a sense, this theorem plays the same role as the corresponding theorem in LP: it limits the search for the optimum to the extreme points, (i.e., extremal flows). As in LP, complete enumeration of extreme points is out of all practical considerations. The objective of analysis, then, is to characterize the extremal flow to a sufficient degree that an efficient algorithm can be constructed. We demonstrate the approach by application to a couple of examples.

First, the following terminology and set of properties are useful in subsequent discussion. Two chains E_1 and E_2 are said to be <u>separate</u> if they have at most one node in common. It immediately follows that in single-source single-terminal flow networks, there cannot be two separate chains from source to terminal. On the other hand, a network with K sources can have at most K separate chains from source to terminal. And a network with K sources can have at most K separate chains to any node (called K <u>source chains</u>) unless the node in question is itself a source and then there can be no more than K-1 separate source chains. This, in turn, leads to the conclusion that if node $\underline{j}$ has K separate source chains then $\underline{j}$ can have at most K positive inputs. For otherwise one can trace back each chain to at least one other common point (besides $\underline{j}$ itself).

An immediate corollary of the above is that in a network with K sources, the existence of extremal flow implies that each node $\underline{i} \varepsilon N$ has at most K positive inputs, unless $\underline{i}$ itself is a source and then the number is reduced to K-1. In other words, extremal flow implies separateness, and the corollary follows.

The notion of separateness is also useful in single source networks, whose extremal flow can be further characterized to advantage. In particular, we shall conclude that an extremal flow is an arborescence flow, which will immediately lead to efficient algorithms, usually of the dynamic programming type.

To this end, let us define an s-t chain as a source to t_j chain to mean that it is a chain from some source $\underline{s}$ to terminal $\underline{t}_j$. Notice that the chain is distinguished by its terminal (not originating or source) node. Two chains E_1 and E_2 may have a node $\underline{e}$ in common other than the source, see Fig. 2.19a. (In that case, they are not separate.) A collection D of such s-t chains is called an

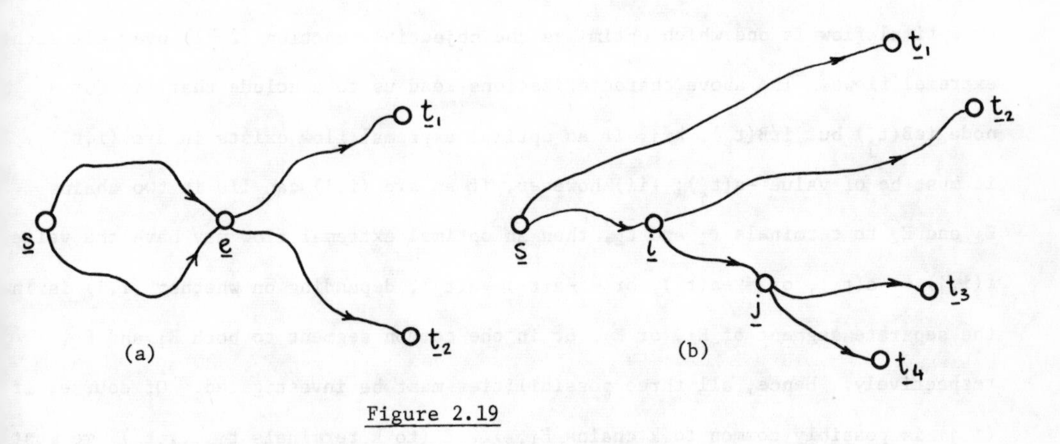

Figure 2.19

arborescence if the following two conditions hold:

1. There is one and only one s-t chain in the collection terminating at node $\underline{t}_j$, j=1,...,K
2. If node $\underline{e}$ is in two s-t chains, then the two chains coincide prior to $\underline{e}$.

An example of an arborescence is schematically shown in Fig. 2.19b. A flow f is called an <u>arborescence flow</u> if it is feasible and $f(i,j) > 0$ on the arcs (i,j) of each s-t chain in the arborescence and $f(i,j) = 0$ on all other arcs.

As the reader may have already suspected, an extremal flow is an arborescence flow, and conversely. The proof of this assertion follows easily from the definition of both. Its implication is obvious: it characterizes further an extremal flow, thus narrowing the domain of search for the optimal flow to arborescence flow. But such flow has the following two properties:

1. If an $s-t_j$ chain is separate (i.e., it shares no other node but the source <u>s</u> with any other chain; see chain $s-t_1$ of Fig. 2.18b) then it must be true that the flow in each arc of that chain is equal to $-a(t_j) > 0$.
2. If a node <u>e</u> lies in m chains to terminals $t_1,\ldots,t_m$ then the flow into <u>e</u> must be equal to $-\sum_{i=1}^{m} a(ti) > 0$ (for example, the flow into node <u>j</u> of Fig. 2.19b must be equal to $-[a(t_3) + a(t_4)]$).

Though these two properties can be rigorously proven, a mere glance at an arborescence flow will convince the reader of their validity.

We now have a more comprehensive description of an extremal flow. Naturally an optimal flow is one which optimizes the objective function (2-32) over all such extremal flows. The above characterizations lead us to conclude that: (i) for a node $\underline{i} \varepsilon B(t_j)$ but $i \notin B(t_k)$, $k \neq j$, if an optimal extremal flow exists in arc (i,t_j) it must be of value $-a(t_j)$; (ii) however, if an arc (i,j) can lie in two chains E_1 and E_2 to terminals t_1 and t_2, then an optimal extremal flow may have the value $f(i,j) = -a(t_1)$, or $= -a(t_2)$, or $= -a(t_1) - a(t_2)$, depending on whether (i,j) is in the separate segment of E_1, or E_2, or in the common segment to both E_1 and E_2, respectively. Hence, all three possibilities must be investigated. Of course, if (i,j) is possibly common to k chains $E_1, \ldots, E_k$ (to k terminals $t_1,\ldots,t_k$), we must investigate all 2^k-1 possibilities: $f(i,j) = -a(t_1), -a(t_2),\ldots, -a(t_k)$; $[-a(t_1) - a(t_2)],\ldots, [-a(t_{k-1}) - a(t_k)]; \ldots, -\sum_{i=1}^{k} a(t_i)$. This suggests a dynamic programming approach. We exhibit the formulation for the case of two terminals

t_1 and t_2:

Let $v_1(i)$ be the minimum cost of shipping $-a(t_1)$ units from node $\underline{i}$ to terminal node t_1; let $v_2(i)$ be defined similarly but with $-a(t_2)$; and let $v_{1,2}(i)$ be the minimum cost of shipping $-a(t_1)$ $-a(t_2)$ units from node $\underline{i}$ with $-a(t_1)$ destined to terminal t_1 and $-a(t_2)$ destined to terminal t_2. Then we have, in typical dynamic programming fashion,

$$v_1(i) = \min_{j \in A(i)} \{c_{ij}(-a(t_1)) + v_1(j)\} \quad , \quad \text{for all } i \neq t_1 \tag{2-35a}$$

$$v_2(i) = \min_{j \in A(i)} \{c_{ij}(-a(t_2)) + v_2(j)\} \quad , \quad \text{for all } i \neq t_2 \tag{2-35b}$$

and

$$v_{1,2}(i) = \min_{j,k \in A(i)} \{c_{ij}(-a(t_1) - a(t_2)) + v_{1,2}(j);\ c_{ij}(-a(t_1)) + v_1(j)\ ;\ c_{ik}(-a(t_2)) + v_2(k)\} \tag{2-35c}$$

where $v_1(t_1) = v_2(t_2) = 0$, $v_{1,2}(t_1) = v_2(t_1)$, and $v_{1,2}(t_2) = v_1(t_2)$. The relations (2-35) are solved recursively starting with t_1 and t_2 and moving 'backwards' towards the source, until $v_{1,2}(s)$ is evaluated.

Admittedly, the DP formulation becomes unwieldy for more than three terminals. This points out that research should be directed towards an even closer characterization of the optimal extremal flow.

§2.8 SOME APPLICATIONS

(i) The Warehousing Problem with Concave Costs

The warehousing problem was stated in the Introduction, §2.1, and its solution in the case of linear costs was given in §2.6. In the case of concave procurement costs $Q_i(q_i)$, warehousing costs $W_i(w_i)$ and holding costs $H_i(h_i)$, as well as convex selling values $S_i(s_i)$ the objective can be stated as:

$$\text{minimize} \quad \sum_{i=1}^{T} [Q_i(q_i) + W_i(w_i) + H_i(h_i) - S_i(s_i)]. \tag{2-36}$$

Obviously, the constraints are the same as Eqs. (2-31) and the network representation is identical to Fig. 2.16. Since this is a single-source single-terminal concave cost network, we know that the optimal flow is along one (separate) chain from the source to the terminal. The flow throughout the chain is exactly equal to c, the warehouse capacity. At that value of flow, one can easily evaluate the cost along all the arcs of the network. Equally obvious, the optimal chain is the minimal cost chain; i.e., the shortest path from source to terminal if we interpret the costs as lengths, and this latter is certainly an easy problem to solve.

(ii) A Location-Allocation Problem

We recall the statement of this problem. A company contemplates selling a commodity in n markets; where the demand at location j is equal to r_j; j=1,...n. To satisfy such demand, the company has selected m cities as the best candidates for plant locations, but the number of plants has not been determined yet. In other words, if one plant is built, it must satisfy all the demand; if two or more plants are built (but no more than m), the company must decide on where to build them (the location problem) and on the territory to be served by each plant (the allocation problem). Of course, any given allocation determines the productive capacity of each plant. The question is: how many plants should be built ($\leq m$), and what allocation (of markets to plants) should be made to maximize the total discounted income over a planning horizon of L years? The cost parameters are as follows. If plant i produces X_i units annually, the discounted cost over L is a concave function $C_i(x_i)$; if w_{ij} units are shipped from plant i to demand center j, the discounted cost over L is $W_{ij}(w_{ij})$ assumed also concave; and the income from such shipments is $S_{ij}(w_{ij})$, assumed convex. Then, it is desired to:

$$\text{minimize} \quad \sum_{i=1}^{m} C_i(x_i) + \sum_{i,j} [W_{ij}(w_{ij}) - S_{ij}(w_{ij})] \tag{2-37}$$

such that

$$x_i - \sum_{j=1}^{n} w_{ij} = 0 \qquad i=1,\ldots,m \tag{2-38a}$$

$$\sum_{i=1}^{m} w_{ij} = -r_j \qquad j=1,\ldots,n \tag{2-38b}$$

$$\sum_{i=1}^{m} x_i = \sum_{j=1}^{n} r_j \tag{2-38c}$$

and $x_i \geq 0$, $w_{ij} \geq 0$. This is a single commodity single source multiple destination concave flow-network problem, whose solution is obtained through a general DP formulation similar to Eq. (2-35), (in spite of its computational complexity). For m=2 and n=2 the network is depicted in Fig. 2.20.

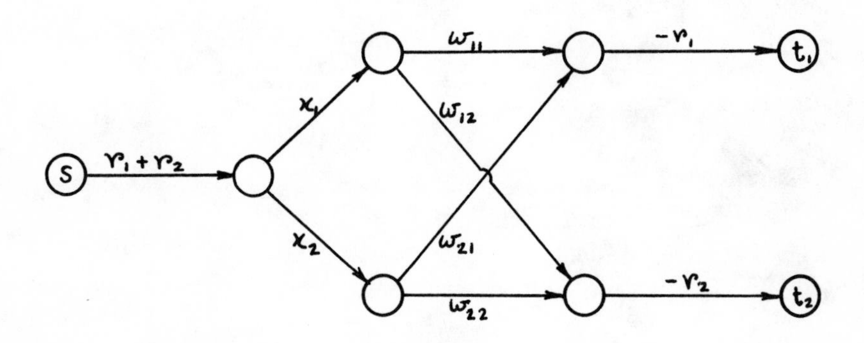

Figure 2.20

As a footnote to this problem, any allocation of n destinations to m sources can be given by a matrix $[\alpha_{ij}]$ with $\alpha_{ij} = 1$ if destination j is allocated to source i and, $\alpha_{ij} = 0$ otherwise. In general there are

$$S(n,m) = \frac{1}{m!} \sum_{k=0}^{m} \binom{m}{k} (-1)^k (m-k)^n$$

different allocation matrices, assuming all sources are indistinguishable with respect to product, capacity, service, etc. An enumerative approach to the problem would study the allocation problem for each of the S(n,m) locations, and then choose the optimal location-allocation pattern. The network flow approach solves for the optimal location-allocation directly.

REFERENCES TO CHAPTER 2

[1] Dantzig, G., "Linear Programming and Extensions, Princeton University Press, 1963.

[2] ____________ and P. Wolfe, "Decomposition Principle for Linear Programs", Opns. Res., Vol. 8, No. 1, 1960.

[3] Eggleston, H. G., "Convexity", Cambridge Tracts in Mathematics and Mathematical Physics No. 47, Cambridge University Press, 1963.

[4] Ford, L. R. and D. R. Fulkerson, "Flows in Networks", Princeton University Press, 1962.

[5] ____________, ________________, "Suggested Computation for Maximal Multi-Commodity Network Flows", Mgt. Sc., Vol. 5, No. 1, 1958.

[6] Fulkerson, D. R., "An Out-of-Kilter Method for Minimal Cost Flow Problems", Jour. of SIAM, Vol. 9, No. 1, 1961.

[7] ________________, "Flow Networks and Combinatorial Operations Research", Amer. Math. Monthly, Vol. 73, 1966, pp. 115-138.

[8] Gomory, R. E. and T. C. Hu, "Multi-Terminal Network Flows", Jour. of SIAM, Vol. 9, No. 4, 1961.

[9] Hoffman, A. J., "Some Recent Applications of the Theory of Linear Inequalities to Extremal Combinatorial Analysis", Proc. Symposia on Applied Math., Vol. 10, 1960.

[10] Hu, T. C., "Recent Advances in Network Flows", Math. Res. Center, Tech. Summary Report #753, June 1967, The University of Wisconsin.

[11] Klein, M., "A Primal Method for Minimal Cost Flows with Applications to the Assignment and Transportation Problems", Mgt.Sc., Vol.14, No.3, 1967.

[12] Zangwill, W. I., "Minimum Concave Cost Flows in Certain Networks", Working Paper, Center for Research in Mgt. Sc., University of California, July 1967.

CHAPTER 3

SIGNAL FLOW GRAPHS

Contents | Page

§3.1 INTRODUCTION AND TERMINOLOGY

Signal Flow Graphs (SFGs), sometimes referred to simply as 'flowgraphs', are an analytic tool often used in the modeling and analysis of linear systems. Although their use originated in the analysis of electrical networks, increased interest in SFGs derives from the importance of the analysis and synthesis of linear systems occurring in many fields of science and engineering. It is a known fact that, irrespective of their content, many systems can be modeled as a set of linear equations, to which the methodology of SFG theory is directly applicable. As we shall see below, such diverse problems as stochastic systems of the Markovian or semi-Markovian types, as well as dynamic systems giving rise to linear differential equations, can be analyzed through SFGs.†

The generic element of signal flow graphs (SFGs) is shown in Fig. 3-1. It consists of a directed arc (i.e., an arrow) connecting two nodes, the 'origin' x_1 and the 'terminal' x_2. The arc is said to be of 'transmittance' t_{12} meaning that the value of the node x_1 is multiplied by t_{12} as it is transmitted through the arc such that the

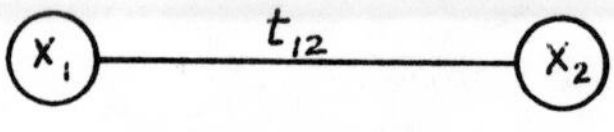

x_1 —— t_{12} —— x_2

Figure 3-1

† As a sample of other applications see refs. [4] and [6].

variable represented by the node x_2 is defined by

$$x_2 = x_1 t_{12}.$$

In general, more than one arc may leave any node and more than one arc may terminate at any node. In the former case, the value of the node is multiplied by the transmittance of each arc emanating from it; while in the latter case, the value of the node is equal to the sum of all the inputs into it. These two properties define the 'equivalent transmittance' in the case of parallel paths, see Fig. 3-2:

$$x_2 = x_1(t_{12} + t'_{12} + t''_{12}).$$

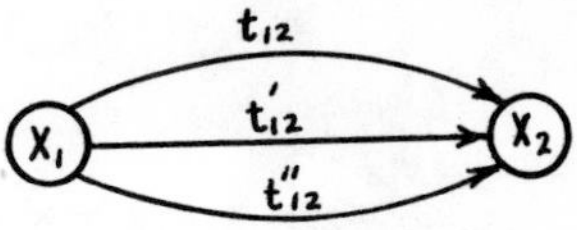

Parallel Paths

Figure 3-2

Now, in the analysis of signal flow graphs one is usually interested in the transmittance between a selected subset of nodes. The graphic-algebraic procedure utilized to reduce a SFG to a residual graph showing only the nodes of interest is based on a few simple rules. These rules are derived from the fact that a SFG is basically a representation of a set of linear equations. And it is this very fact that bestows importance on SFG theory because the behavior of so many physical systems can be described by linear algebraic equations, as will be amply demonstrated below.

Instead of cluttering this manuscript with a long list of such rules for graph reduction, and with proofs of their validity, we shall content ourselves with constructing the minimal required set so that we can proceed with the more interesting (and much more rewarding) job of applications.

We have already specified the rule for combining arcs in parallel. On the other hand, it must be obvious that for two arcs in series, as in Fig. 3-3, the value of x_3 is given by $x_3 = x_2 t_{23} = x_1 t_{12} t_{23}$. From this and the previous rule, we conclude that: transmittances of arcs in series

Branches in Series

Figure 3-3

multiply and of arcs in parallel add.

We shall oftentimes be concerned with feedback loops, as shown in Fig. 3-4a, and the question is: what is the value of the variable x_4? We develop the general rule as follows. From the graphs we have

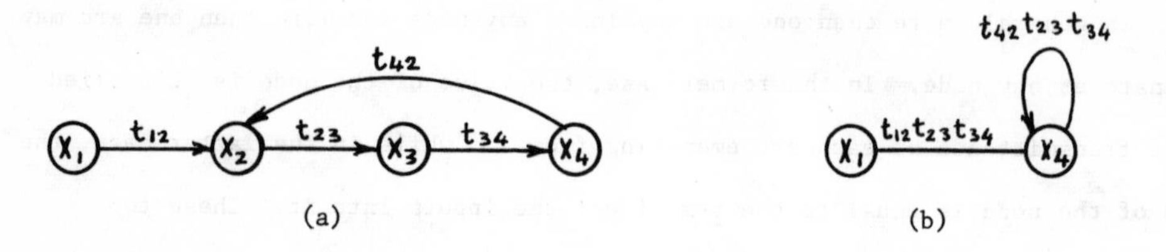

Feedback and Self-Loops

Figure 3-4

$$x_2 = x_1 t_{12} + x_4 t_{42}$$

$$x_3 = x_2 t_{23}, \text{ which upon substitution yields}$$

$$= x_1 t_{12} t_{23} + x_4 t_{42} t_{23}$$

and finally

$$x_4 = x_3 t_{34}$$

$$= x_1 t_{12} t_{23} t_{34} + x_4 t_{42} t_{23} t_{34}.$$

Notice that this last equation has the SFG representation of Fig. 3-4b. It is then obvious that the transmittance from x_1 to x_4 with the feedback loop of (a) is equivalent to the transmittance of the path π_{x_1,x_4} and the self-loop around x_4 of transmittance equal to the transmittance of the feedback loop of (b). Finally, the last equation reduces to

$$x_4 = x_1 \cdot \frac{t_{12} t_{23} t_{34}}{1 - t_{42} t_{23} t_{34}}$$

which implies that a self-loop of transmittance ℓ is equivalent to $1/(1-\ell)$ multiplied by the transmittance of all the arrows incident into the node.

Although the above three rules (paths in parallel, paths in series and feedback or self-loops) are the cornerstone rules by which one can reduce any SFG to its equivalent residual graph, the reduction of complicated graphs can be rather cumbersome and prone to error. For such graphs Mason[7] proposed a rule for reduction which, in spite of its formidable appearance, is rather simple and straightforward.

In a network composed of several sources, sinks and intermediate nodes†, the transmittance from a source r to a sink s is evaluated as follows:

Π_j denote the transmittance along the jth forward path between the designated source and terminal;

$L_1, L_2, \ldots, L_n$ represent the transmittance of the loops (both feedback and self-loops) in the graph;

$D = 1 - \sum_i L_i + \sum_{i,j} L_i L_j - \sum_{i,j,k} L_i L_j L_k + \ldots$

where each multiple summation extends over non-adjacent loops (two loops are adjacent if they share at least one node).

Δ_j = evaluated as D but with path π_j removed (which implies the removal of all nodes and branches along π_j).

Then the desired transmittance $t_{r,s}$ is given by

$$t_{r,s} = \sum_j \Pi_j \Delta_j / D \qquad (3\text{-}1)$$

With this seemingly meager formal structure we propose to analyze the dynamic behavior of systems in the areas of inventory, queuing, project management and production. But before we indulge in this fascinating and totally absorbing tour it is advisable that the reader be reminded, once again, that the common feature of all these analyses is that the systems in question can be represented by a set of linear algebraic equations. In some cases these algebraic equations are

---

† By convention, a sink is a node with no arcs leaving it, and a source is a node with no arcs entering it. An SFG may have any number of sources and sinks but must have at least one of each.

arrived at by applying transform methods (Fourier, Laplace, probability generating, and z-transforms) to the original equations representing the dynamic behavior of the system. But once such transformation is accomplished, analysis proceeds on the basis of the SFG.

§3.2 AN INVENTORY PROBLEM

Suppose that the manager of a warehouse knows that the demand per period for his product has the following probability distribution:

demand, x	0	1	2	3
prob. of demand, f(x)	.1	.4	.3	.2

Being of a conservative nature, he decides not to replenish his stock unless he ends a period with nothing 'on hand', and then he would order 3 units. His ordering cost is $1.0 per replenishment; his carrying cost os $0.1 per unit in stock at the end of a period; and he profits $3.0 from every unit sold (this is the difference between his sale price and the purchase price (or 'cost of product sold') of the unit). The 'lead time' for replenishment is one period; i.e., if at the end of a period n he places an order (for 3 units), the stock will not be replenished except at the <u>end</u> of period n + 1 (or equivalently, at the beginning of period n + 2). What can be said about this inventory system?

Obviously, at the beginning of any period n he can be in any one of four states: x_0, which signifies that he has nothing in stock; x_1, which signifies that he has one unit 'on hand'; x_2 and x_3, which are defined similarly. During any period he can 'move' to any one of the four states and end the period (or, equivalently, start period n + 1) in one of the four states. The 'law' governing his 'movement' from state x_i to state x_j is clearly the probability law of demand. For instance, if he is in x_3 at the beginning of the period, he would 'move' to state x_3 with probability $p_{3,3} = f(0) = 0.1$; namely, if nothing is demanded, which occurs with probability 0.1. Similarly, he would 'move' to x_2 if exactly one unit is sold, i.e., with probability $p_{3,2} = f(1) = 0.4$ and he would 'move' to x_1 with probability $p_{3,1} = f(2) = 0.3$; and he would move to x_0 with probability $p_{3,0} = f(3) = 0.2$. A similar analysis can be applied to his other three starting states, and the 'transition' probabilities can be represented in matrix format as follows:

Terminal State x_j

Starting State x_i	0	1	2	3
0				1
1	.9	.1		
2	.5	.4	.1	
3	.2	.3	.4	.1

This is the well-known stochastic matrix of transition probabilities. The inventory process is a Markov chain because the probability of transition is dependent on the current state x_i (and perhaps the future state x_j), and is independent of all past history (or how the system got to state x_i). In other words, it is a memoryless process.

It is also a renewal process, because 'life starts afresh' every time the 'on hand' inventory hits the zero level, or, for that matter, if the system is at any state whatsoever.

Introducing: the z-transform

We know perfectly well that the duration of a transition is exactly one period. Suppose, for the moment, that this were not the case (as, in fact, it is not in semi-Markov processes, to be dealt with in greater detail below). Let $h_{ij}(\tau)$ denote the probability mass function of the duration of the transfer from x_i to x_j, $\tau = 1, 2, \ldots$ Let $h_{ij}(z)$ be the z-transform of $h_{ij}(\tau)$, defined by

$$h_{ij}(z) = \sum_{\tau} h_{ij}(\tau)z^{\tau}, \quad \tau \geq 0 \tag{3-2}$$

In the case of discrete probability measures, $h_{ij}(z)$ is identical with the probability generating function of h_{ij}. The advantages of this transformation are numerous, not the least important of which is the fact that convolutions in the time domain are replaced by multiplications in the z-domain. For instance, the probability that it takes exactly t time units to transfer from state x_i to state x_k via state x_j is given by

$$h_{ik}(t) = \sum_{\tau=0}^{t} h_{ij}(\tau)\, h_{jk}(t-\tau)$$

It is easy to show that

$$h_{ik}(z) = h_{ij}(z) h_{jk}(z).$$

From this we conclude that if we interpret $p_{ij}h_{ij}(z)$ as the 'transmittance' between nodes x_i and x_j, the transmittance of two arcs in series is precisely the product of the two transmittances. This is only half of the justification for the use of SFGs; the other half is the arcs-in-parallel case.

Suppose that the system can transfer from x_i to x_j via two routes: the first with 'holding time' of probability mass $h_{ij}^{(1)}(\tau)$ and the second of probability mass $h_{ij}^{(2)}(\tau)$. Suppose further, that the system chooses between the two routes with probabilities $p_{ij}^{(1)}$ and $p_{ij}^{(2)}$ (if these are the only two transfer routes from x_i, then clearly $p_{ij}^{(1)} + p_{ij}^{(2)} = 1$). Obviously, the equivalent probability mass function of transfer from x_i to x_j is simply

$$p_{ij}^{(1)} h_{ij}^{(1)}(\tau) + p_{ij}^{(2)} h_{ij}^{(2)}(\tau)$$

whose z-transform is

$$p_{ij}^{(1)} h_{ij}^{(1)}(z) + p_{ij}^{(2)} h_{ij}^{(2)}(z) .$$

Hence we have verified that transmittances in parallel add, which completes the proof that $p_{ij}h_{ij}(z)$ can be used as the transmittance between x_i and x_j.

Transient and Steady-State Probabilities

We return now to our inventory problem and specialize $h_{ij}(\tau)$ to the degenerate case $h_{ij}(\tau) = 1$ if $\tau = 1$ and $= 0$ otherwise. Hence $h_{ij}(z) = z$ for all i and j. The SFG of the system is as shown in Fig. 3-5, in which we assume the system starts in state x_3.

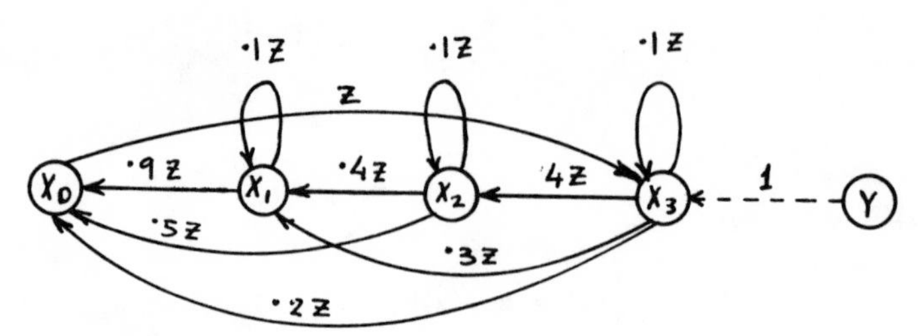

Signal Flow Graph of Inventory Problem

Figure 3-5

Application of Mason's rule, Eq. (3-1), yields,

$$D(z)^{\dagger} = 1 - .3z - .17z^2 - .431z^3 - .099z^4$$

$$t_{y,x_3} = \frac{1 - .2z + .01z^2}{D(z)}$$

$$t_{y,x_2} = \frac{.4z(1 - .1z)}{D(z)}$$

$$t_{y,x_1} = \frac{z(.3 + .13z)}{D(z)}$$

$$t_{y,x_0} = \frac{z(.2 + .43z + .099z^2)}{D(z)}$$

As a check,

$$\sum_{i=0}^{3} t_{y,x_i} = \frac{1 + .7z + .53z^2 + .099z^3}{D(z)} = \frac{1}{1-z} = 1 + z + z^2 + z^3 + \ldots$$

† We remark that $D(1) = 0$, which is always true for recurrent Markov processes, and provides an easy check on numerical calculations.

which is an eminently logical result since the probability of being in one of these four states is 1 for all time.

Of special interest to the manager is the transmittance T_{y,x_0} because it indicates his out-of-stock condition. By long division we see that

$$t_{y,x_0} = .2z + .49z^2 + .28z^3 + .2533z^4 + .35329z^5 + \ldots$$

The coefficient of z^n is, by definition, the probability that the system which started in state x_3 will be in state x_0 at time n. It can be seen that the probability stabilizes to a value around 0.3. Let us denote this steady state probability by p_0. The value of p_0 can be determined to any desired degree of accuracy through carrying out the long division to as many terms as deemed necessary. This is tedious. Consider, as an alternative, the following reasoning. If p_0 is the steady state value of the coefficient of z^n as $n \to \infty$, then, except for an error that approaches 0 as $n \to \infty$, the above series expansion is equal to $p_0 \sum_{n=0}^{\infty} z^n = p_0/(1-z)$. Hence

$$p_0 = \underset{z \to 1}{\text{limit}}\ (1-z)t_{y,x_0}$$

which is the well-known 'final value theorem' in the theory of z-transforms. Application to our expression yields

$$p_0 = \lim_{z \to 1} (1-z) \frac{z(.20 + .43z + .099z^2)}{D(z)}$$

$$= \lim_{z \to 1} \frac{z(.20 + .43z + .099z^2)}{1 + .7z + .53z^2 + .099z^3} = \frac{.729}{2.329} \simeq .313$$

In a similar fashion we deduce that

$$p_1 = \lim_{z\to 1} \frac{z(.30 + .13z)}{1 + .7z + .53z^2 + .099z^3} = \frac{.43}{2.329} \simeq .18462$$

$$p_2 = \lim_{z\to 1} \frac{.4z(1 - .1z)}{1 + .7z + .53z^2 + .099z^3} = \frac{.36}{2.329} \simeq .1112$$

$$p_3 = \lim_{z\to 1} \frac{1 - .2z + .01z^2}{1 + .7z + .53z^2 + .099z^3} = \frac{.81}{2.329} \simeq .39118$$

The reader can easily verify that these steady state probabilities are independent of the initial state of the system (that is why they were given one subscript only). For example, if the initial position of the system were x_2 (which can be represented on the SFG of Fig. 3-5 by letting the arc of transmittance 1 from node y be incident on node x_2 instead of x_3), we deduce that

$$(t_{y,x_1} \,|\, \text{system started at } x_2) = \frac{.4z(1 - .1z - .2z^2) + .15z^3}{D(z)}$$

$$= \frac{z(.4 - .04z + .07z^2)}{D(z)}$$

such that

$$p_1 = \lim_{z\to 1} (t_{y,x_2} \,|\, \text{system started at } x_2) = \frac{.43}{2.329}, \text{ as before.}$$

Of course, the difference between the two expressions of t_{y,x_1} lies in the transient behavior of the system, and is understandably different for the two different starting positions.

The reader will notice that we have just demonstrated, though we have not formally proved, the well-known theorem of finite-state irreducible discrete-time Markov processes: that a stationary probability mass function exists for the states of the system, and is independent of the initial position of the system.

Some Cost Considerations

We proceed with our analysis. The expected cost of this policy is

$$\underbrace{\$1\times.313}_{\text{ordering}} + \underbrace{\$.1(.18462 + 2\times.1112 + 3\times.39118)}_{\text{inventory}} = \$.471.$$

The expected income is

$$\underbrace{\$3(1\times.18462\times.9)}_{\text{in state } x_1} + \underbrace{.1112(1\times.4 + 2\times.5)}_{\text{in state } x_2} + \underbrace{.39118(1\times.4 + 2\times.3 + 3\times.2)}_{\text{in state } x_3}$$

$$= \$.947726,$$

so that his expected net profit is $\simeq$ \$.4767 per period.

In the statement of the problem no cost was attached to being out of stock. In other words, it was assumed that demand beyond what is available 'on hand' is lost, and that is the end of that. But it is not wholly illogical to accord some cost to out-of-stock conditions. This can be a 'lump sum' cost incurred whenever customers are denied service, or it may be a cost proportional to the quantity demanded but not provided, or both.

For instance, whenever x_0 is reached at the end of any period, the manager is sure of turning away his customers for one whole period. If he subjectively evaluates his loss at 10% of the average possible gain, he should subtract from the net profit figure evaluated above his expected loss due to this condition, which is equal to

$$.1 \times \$3 \left[\sum_{x=0}^{3} x f(x)\right] \times p_0 \simeq \$.15.$$

This reduces his net expected profit down to $\simeq$ \$.3267 per period.

Can he do better? This is a very natural and fundamental question, which, incidentally, transcends analysis and moves into synthesis, because its logical extension is: what is his best? Obviously, if our store manager is to do better, he must change his decision concerning replenishment.† But this would change the structure of the SFG! Translated, therefore, the question of optimality is ultimately related to the determination of the optimal structure of the SFG. This should come as no surprise to anyone who has had any experience with design problems of either hardware or conceptual systems. Typically, such problems are orders of magnitude more difficult than their corresponding analysis problems. It turns out that this particular problem can be solved by the Howard[5] algorithm of dynamic programming. We shall not pursue this point any further or else it would take us too far afield from our main theme.

Some Further Questions of Analysis

For any given replenishment policy adopted by our manager there are several interesting and meaningful questions of analysis which are related to the transient or the steady state conditions of the system. We now consider a few of these questions.

1. Duration of first passage through any state x_j when the system starts in state x_i.

This is easily obtained by removing all arcs out of state x_j and then evaluating the transmittance t_{ij}. This yields the probability generating function of the duration of the first passage.

For example, if we wish to find the duration of first passage through x_0 when the system starts at x_3, we eliminate the arc x_0x_3 and evaluate

$$t_{x_3,x_0} = \frac{z(.2 + .43z + .099z^2)}{1 - .3z + .03z^2 - .001z^3}$$

The probability that the system will ever be in state x_0 is simply given by

$$t_{x_3,x_0}\Big|_{z=1} = 1.0.$$

---

† We are excluding from the discussion other strategies open to him such as more advertising, improved product, diversification to other products, etc.

Hence, x_0 is a certain event, as is well known. By the definition of the z-transform (or probability generating function), $t_{x_3,x_0} = \sum_{n=0}^{\infty} p_{3,0}^{(n)} z^n$; hence the various moments can be obtained by differentiation. For example,

$$\text{mean duration to first passage} = \frac{d}{dz} t_{x_3,x_0}\Big|_{z=1} = 2.19 \text{ periods,}$$

and

$$\text{variance of duration to first passage} = \frac{d^2}{dz^2} t_{x_3,x_0}\Big|_{z=1} + \frac{d}{dz} t_{x_3,x_0}\Big|_{z=1}$$

$$- \left(\frac{d}{dz} t_{x_3,x_0}\Big|_{z=1}\right)^2 = \frac{1.7982}{.53144} + 2.19 - (2.19)^2 \simeq .7613 \text{ (period)}^2$$

or a standard deviation of .8725 periods.

Incidentally, the limiting value $\lim_{z\to 1} (1-z)t_{x_3,x_0} = 0$, which implies that the probability of first passage $\to 0$ as $n \to \infty$; again an intuitively obvious result.

We also draw the reader's attention to the fact that the SFG of Fig. 3-5 with arc x_0x_3 removed is no more a graph of a Markov process because the probability out of state x_0 is 0. This can be remedied, however, by having a self-loop on x_0 of transmittance z which would indicate that x_0 is an absorbing state: once it is entered the probability is 1 that the system will remain there.

Clearly, a similar approach can be applied to any other node to obtain the probability generating function of first passage to that node.

2. Duration of kth passage through any state x_j.

Of course, first passage is a special case of the kth passage, and maybe we should have discussed this general question first and then specialized the result to the first passage. In any event, what is the distribution of the second passage time through x_1, say?

The answer is easily achieved through 'tagging' the arcs into the node of interest, x_1 in this case, with r, as shown in Fig. 3-6. Now, every time the

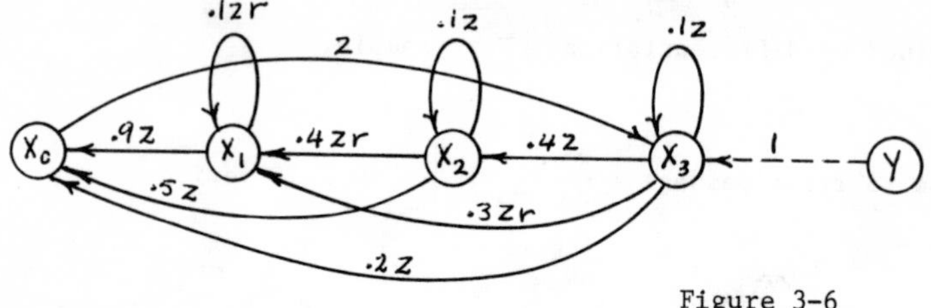

Figure 3-6

system transfers into x_1 an r will be 'picked up', and if x_1 is visited k times we must look for the coefficient of r^k in the expansion of $t_{3,1}(z,r)$.

Let us take a closer look at such an expansion:

$$t_{3,1}(z,r) = a_1(z)r + a_2(z)r^2 + \ldots + a_k(z)r^k + \ldots$$

in which the coefficient $a_k(z)$ represents the conditional z-transform that node x_1 has been passed exactly k times. In other words, $a_k(z)$ is the z-transform of the combined event: node x_1 is realized and k entries into x_1 have occurred. The function a(z) is equal to the product of the probability, p_k that x_1 is visited k times and the conditional probability generating function $M_k(2)$;$a_k(z) = p_k\ M_k(z)$; whence $p_k = a_k(1)$. Consequently, the conditional pgf $M_k(z) = a_k(z)/a_k(1)$. The conditional $a_k(z)$ can be obtained from the transmittance $t_{3,1}(z,r)$ as

$$a_k(z) = \frac{1}{k!}\ \frac{d^k}{dr^k}\ t_{3,1}(2,r)\ \Bigg|_{r=0}\ .$$

Consequently, the conditional probability generating function is

$$(t_{3,1}(z)\ |\ x_1\text{:k times}) = \frac{d^k}{dr^k}\ t_{3\ 1}(z,r)\quad {}_{r=0}\quad \frac{d^k}{dr^k}\ t_{3\ 1}(z,r)\ \Bigg|_{\substack{r=0\\z=1}} \qquad (3\text{-}3)$$

Application to Fig. 3-6 yields:

$$t_{3,1}(z,r) = \frac{zr(.3 + .13z)}{(1-.2z-.19z^2-.18z^3) - r(.1z-.02z^2+.251z^3+.099z^4)}$$

$$= \frac{z(.3+.13z)}{B}\left[r + \frac{Ar^2}{B} + \frac{A^2}{B^2}r^3 + \dots\right]$$

where $A = .1z - .02z^2 + .251z^3 + .099z^4$

$B = 1 - .2z - .19z^2 - .18z^3$.

Because of the simple form of $t_{3,1}(z,r)$ it was possible to expand it as a series in powers of r. We immediately deduce

$$\left.\begin{array}{l}\text{probability generating function}\\ \text{of being in state } x_1\text{, given that}\\ x_1 \text{ has been visited twice.}\end{array}\right\} = \frac{z(.3 + .13z)A}{B^2}$$

In this particular instance (as it is generally true for recurrent Markov systems), the normalizing factor is 1 (because, with probability 1 state x_1 will be visited twice or, for that matter, k times).

Now that we possess the pgf, we can deduce the moments of the conditional second passage by successive differentiation in the usual fashion. For example, the average duration to the second passage is given by

$$\left.\frac{d}{dz}\,\frac{[z(.3 + .13z)A]}{B^2}\right|_{z=1} = 9.323 \text{ periods.}$$

<u>Some Counting Problems</u>.

Signal flow graphs, by their very pictorial nature, provide simple and intuitively appealing devices for 'counting' events, such as the number of times a node is realized given that another node is realized (or not realized); the number of times the system is in a particular state before transferring to an absorbing state; etc.

For example, suppose that in our inventory problem the state x_0 is an absorbing state; in other words, the inventory starts with 3 units and will never be replenished. In such a case, the SFG is similar to Fig. 3-5 with the exception that the 'renewal' path x_0x_3 is removed and is substituted by a self-loop of transmittance z around node x_0. Question: What is the number of times the system will be in state x_1 before 'going out of business'? The answer is obtained by 'tagging' each arc going into node x_1 with the counting tag r. The resultant graph would be similar to Fig. 3-6 except for the absence of the arc x_0x_3 and the presence of the self-loop around x_0. The transmittance

$$t_{3,0}(z,r) = \frac{z(.2 + .18z) + rz^2(.25 + .099z)}{(1 - 1.2z + .21z^2 - .01z^3) - rz(.1 - .12z + .021z^2 - .001z^3)}$$

$$= \left(\frac{z}{1-z}\right) \frac{(.2 + .18z) + rz(.25 + .099z)}{(1 - .1z)^2\ (1 - .1zr)}$$

Notice that $(1 - z)\ t_{3,0}(z,r)\Big|_{\substack{r=1\\z=1}} = 1$, which is a foregone conclusion since the steady state probability of being in state x_0 is 1, being an absorbing state and all other states transient.

Expanding $t_{3,0}(z,r)$ in powers of r would yield a series

$$t_{3,0}(z,r) = b_0(z) + b_1(z)\ r + b_2(z)r^2 + \ldots$$

in which the coefficient $b_k(z)$ is interpreted as the z-transform of the probability that the system transfers from x_3 to x_0 and visits state x_1 exactly k times. Clearly, $(1-z)\ b_k(z)\Big|_{z=1}$ is the probability of that (composite) event. For example,

$$t_{3,0}(z,r) = \left(\frac{z}{(1-z)(1-.1z)^2}\right)\left(\frac{(.2 + .18z) + rz(.25 + .099z)}{(1 - .1zr)}\right)$$

from which we deduce that

$$\Pr(x_3 \to x_0 \text{ without passing by } x_1) = \lim_{z\to 1} \frac{z(.2 + .18z)}{(1 - .1z)^2} = \frac{.38}{(.9)^2}$$

$$\Pr(x_3 \to x_0 \text{ and visits } x_1 \text{ once}) = \lim_{z\to 1} \frac{z}{(1 - .1z)^2} [.1z(.2 + .18z) + z(.25 + .099z)] = \frac{.387}{(.9)^2}$$

$$\Pr(x_3 \to x_0 \text{ and visits } x_1 \text{ twice}) = \lim_{z\to 1} \frac{z}{(1 - .1z)^2} [.01z^2(.2 + .18z) + .1z^2(.25 + .099z)] = \frac{.0387}{(.9)^2}$$

and, in general, $\Pr(x_3 \to x_0 \text{ and visits } x_1 \text{ k times}) = \frac{.387}{(.9)^2} (.1)^{k-1}$. Consequently, the probability of visiting x_1 exactly k times $\to 0$ as $k \to \infty$; the probability of ever visiting x_1 is $\sum_{k=1}^{\infty} \frac{.387}{(.9)^2} (.1)^{k-1} = \frac{.43}{(.9)^2} = 1 -$ (Probability that x_1 will never be visited before absorption at x_0); and the average number of visits to x_1 before absorption is $\sum_{k=1}^{\infty} \frac{.387}{(.9)^2} k\,(.1)^{k-1} = \frac{.387}{(.9)^2} \times \frac{1}{(1 - .1)^2} = .5898$ visits.

Returning to our original renewal process (in which x_0 was not an absorbing state), we notice that the steady state probabilities previously evaluated give us the expected count of visits to any of the 4 states. In other words, after n periods of times we expect to have had 3 units in inventory .39118n times; 2 units in inventory .1112n times, and so forth.

HUNT LIBRARY
CARNEGIE-MELLON UNIVERSITY

§3.3 A QUEUING PROBLEM

Because we wish to draw graphs, we shall confine ourselves to finite queues, an excellent example of which is the well-known "machine minding" problem. This is the problem of an operator (or a few operators) who is responsible for the proper functioning of a number of machines. If a machine fails, he attends to it. Sometimes a machine need not 'fail' in the sense of breaking down; it is sufficient that its operation is halted and it is in need of the attendant's services, such as the need for 'setting up' a turret lathe, or re-feeding a screw-machine, or re-threading a weaving machine, etc. If a second machine fails while the operator is busy with the first machine, it will have to wait. Therefore, the total 'downtime' of the second machine is composed of two parts, the first is the normal need for repair that is to be expected for any piece of machinery of its particular type, and the second is the time wasted because the operator was not available. This latter is caused by the interference of the first machine with the expeditious servicing of the second. This is why the problem is sometimes referred to as the 'machine interference' problem instead of the machine minding problem.

If a third machine breaks down also while the first machine is still being worked on, it will have to wait together with the second machine, and a queue of waiting 'customers' has already been started and is on its way to assert its demands on the operator's time. Of course, such a queue has a maximum length of n machines (including the one being serviced, which is the customary designation of the length of such queues), which by assumption is finite. Needless to say, there are many other queuing situations in which it is legitimate to consider infinite queues. But we shall not occupy ourselves with such systems since, again, this would take us far afield from our main stream of thought.

A final remark is in order before we proceed with the application of SFGs to queuing systems. As soon as q, the number of machines in the queue (which always includes the one being serviced), exceeds 2, there arises the question of priority of service; namely, the order in which the waiting customers will be serviced. This is one more complication which we shall not concern ourselves with here.

In fact, since our purpose is basically to illustrate the applicability of SFGs to the analysis of such systems, we shall consider the simple system composed of one operator and only two machines. We shall assume completely random failure of either machine at the rate $\lambda > 0$ per unit time, which is equivalent to saying that in an infinitesimal period of time Δt the probability of a machine failing is approximately $2\lambda\Delta t$ if both of them were working and $\lambda\Delta t$ if only one of them were working. This, in turn, implies that the probability of non-failure in the same interval Δt is $1 - 2\lambda\Delta t$ and $1 - \lambda\Delta t$, respectively. Furthermore, if a machine is being worked on by the operator, the probability that it will be operative at the end of the period Δt is equal to $\mu\Delta t$, $\mu > 0$; which implies that the probability that it will not be operative in the same interval is $1 - \mu\Delta t$. The probability of both machines failing or being repaired simultaneously in Δt is equal to 0.

Equations of System Dynamics

The standard analysis of this problem proceeds by writing down the difference equations

$$p_0(t + \Delta t) = p_0(t)\ (1 - 2\lambda\Delta t) + p_1(t)\mu\Delta t$$

$$p_1(t + \Delta t) = p_1(t)\ (1 - \lambda\Delta t - \mu\Delta t) + p_0(t)2\lambda\Delta t + p_2(t)\mu\Delta t$$

$$p_2(t + \Delta t) = p_2(t)\ (1 - \mu\Delta t) + p_1(t)\lambda\Delta t$$

where $p_i(x)$ is the probability of the system having i machines inoperative at time x, and each equation represents the manner in which the state i of the l.h.s. can be realized. Now taking the corresponding $p_i(t)$ term to the l.h.s. of each equation and dividing throughout by Δt and letting $\Delta t \to 0$, one obtains the standard difference-differential equations:

$$\frac{d}{dt} p_0(t) = -2\lambda p_0(t) + \mu p_1(t)$$

$$\frac{d}{dt} p_1(t) = -(\lambda + \mu) p_1(t) + 2\lambda p_0(t) + \mu p_2(t)$$

$$\frac{d}{dt} p_2(t) = -\mu p_2(t) + \lambda p_1(t)$$

From the theory of Laplace transforms it is well known that

$$\text{Laplace transform } \mathcal{L}\left(\frac{d}{dt}\right) g(t) = \int_0^\infty \left(\frac{d}{dt} g(t)\right) e^{-st}\, dt = sG(s) - g^+(0),$$

where G(s) is the Laplace transform of g(t). If our system started with <u>both machines working,</u> then $p_0^+(0) = 1$ and $p_1^+(0) = p_2^+(0) = 0$. The Laplace transform of the above equations can be immediately written as

$$P_0(s) = \frac{1}{s} - \frac{2\lambda}{s} P_0(s) + \frac{\mu}{s} P_1(s)$$

$$P_1(s) = -\left(\frac{\lambda+\mu}{s}\right) P_1(s) + \frac{2\lambda}{s} P_0(s) + \frac{\mu}{s} P_2(s)$$

$$P_2(s) = -\frac{\mu}{s} P_2(s) + \frac{\lambda}{s} P_1(s)$$

Now we represent the three variables P_0, P_1 and P_2 by three nodes and the linear equations by the graph of Fig. 3-7.

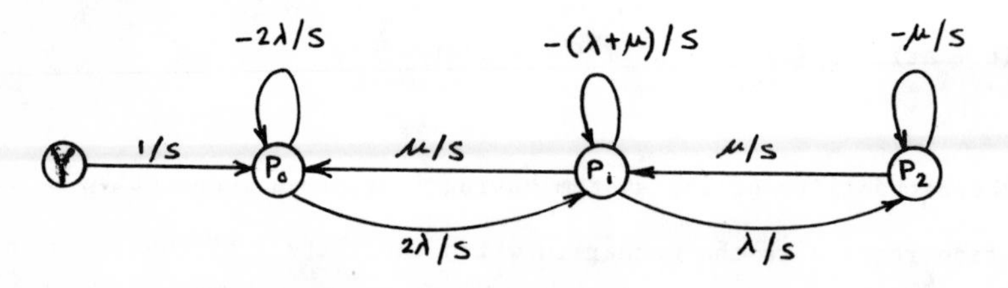

<u>Signal Flow Graph Representation of the</u>

<u>2-Machines, 1-Repairman Problem</u>

<u>Figure 3-7</u>

Transient and Steady-State Probabilities, and Other Considerations

Application of Mason's rule, Eq. (3-1), yields (after normalizing to $\lambda = 1$),

$$D = \frac{1}{s^2} [s^2 + (3 + 2\mu) s + (2 + 2\mu + \mu^2)] = C/s^2$$

where

$$C = s^2 + (3 + 2\mu) s + (2 + 2\mu + \mu^2) = (s - s_1)(s - s_2);$$

$$s_1 = -(3/2 + \mu) + (1/4 + \mu)^{1/2}$$

$$s_2 = -(3/2 + \mu) - (1/4 + \mu)^{1/2} \quad .$$

Hence, if $P_{0,j}(s)$ is the transmittance from node B to node j (see Fig. 3-7), then

$$P_{0,0}(s) = [s^2 + (1 + 2\mu) s + \mu^2]/sC$$

$$P_{0,1}(s) = 2(s + \mu)/sC$$

$$P_{0,2}(s) = 2/sC$$

Expanding each expression into partial fraction (where $R = \sqrt{1/4 + \mu} = \frac{1}{2}(s_1 - s_2)$ and $b = 2 + 2\mu + \mu^2$) yields

$$P_{0,0}(s) = \frac{1}{s}\frac{\mu^2}{b} + \frac{[-.5 + .5\mu + R(1+\mu)]}{2Rb(s - s_1)} + \frac{[.5 - .5\mu + R(1+\mu)]}{2Rbs_2(s - s_2)}$$

$$P_{0,1}(s) = \frac{1}{s}\frac{2\mu}{b} + \frac{-3/2 + R}{R\, s_1(s - s_1)} + \frac{3/2 + R}{R\, s_2(s - s_2)}$$

$$P_{0,2}(s) = \frac{1}{s}\frac{2}{b} + \frac{1}{R\, s_1(s - s_1)} + \frac{-1}{R\, s_2(s - s_2)}$$

In particular, for $\mu = 2 (\lambda = 1)$ we obtain

$$P_{0,0}(s) = \frac{2}{5s} + \frac{1}{3(s+s_1)} + \frac{4}{15(s+s_2)} \Rightarrow \frac{2}{5} + \frac{1}{3} e^{-2t} + \frac{4}{15} e^{-5t} = p_{0,0}(t)$$

$$P_{0,1}(s) = \frac{2}{5s} - \frac{2}{5(s+s_2)} \Rightarrow \frac{2}{5} - \frac{2}{5} e^{-5t} = p_{0,1}(t) \tag{3-4}$$

$$P_{0,2}(s) = \frac{1}{5s} - \frac{1}{3(s-s_1)} + \frac{2}{15(s-s_2)} \Rightarrow \frac{1}{5} - \frac{1}{3} e^{-2t} + \frac{2}{15} e^{-5t} = p_{0,2}(t)$$

where following the arrow we have written the inverse transform in the time domain. By observing the derivative of $p_{0,i}(t)$ we immediately deduce that

$p_{0,0}(t)$ is monotone <u>decreasing</u> from 1 asymptotically to $\frac{2}{5}$

$p_{0,1}(t)$ " " <u>increasing</u> from 0 asymptotically to $\frac{2}{5}$

$p_{0,2}(t)$ " " " " 0 " to $\frac{1}{5}$

This is shown schematically in Fig. 3-8. To be sure, at any time t we have $\sum_{i=0}^{2} p_{0,i}(t) = 1.$

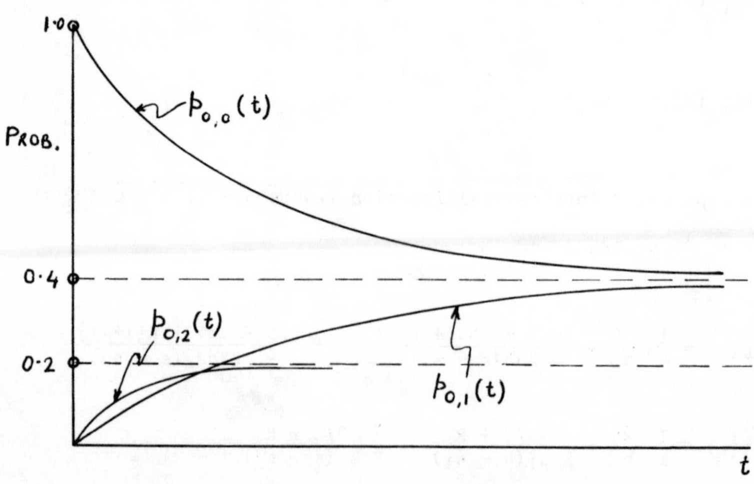

Transient Behavior of the Machine Interference Problem

Figure 3-8

As a corollary to the above evaluations we have that, in the steady state, the mean recurrence time is equal to 2.5 time units for states 0 and 1 and equal to 5 time units for state 2 (i.e., total failure). Moreover, the average length of the queue (i.e., average number of inoperative machines) is 0.8.

Other interesting questions can be answered with equal ease. For instance, what is the probability of continuous operation? Obviously, operation will be discontinued only in the event of both machines failing; hence, the question is asking for the probability of at most one machine failure. To answer this question, we need only eliminate node P_2 and all arcs connected with it. Straightforward analysis of the remaining graph yields

$$P_{0,0}^{(0)}(s) = \frac{s + (1+\mu)}{s^2 + s(3+\mu) + 2} = \frac{a_1}{(s-s_1)} + \frac{a_2}{(s-s_2)}$$

where $$a_1 = (1 + \mu + s_1)/(s_1 - s_2) = (-1 + \mu + \sqrt{1+6\mu+\mu^2})/2\sqrt{1+6\mu+\mu^2}$$

$$a_2 = (1 + \mu + s_2)/(s_2 - s_1) = (1 - \mu + \sqrt{1+6\mu+\mu^2})/2\sqrt{1+6\mu+\mu^2}$$

$$s_1, s_2 = \{-(3 + \mu) \pm \sqrt{1+6\mu+\mu^2}\}/2 \tag{3-5}$$

and

$$P_{0,1}^{(0)}(s) = \frac{2}{s^2 + s(3+\mu) + 2} = \frac{b}{s-s_1} - \frac{b}{s-s_2}$$

where $b = 2/\sqrt{1+6\mu+\mu^2}$. At the particular value $\mu = 2$, we obtain the inverse transforms

$$p_{0,0}^{(0)}(t) = .6212676\ e^{-.438445t} + .3787324\ e^{-4.561555t}$$

$$p_{0,1}^{(0)}(t) = .4850707\ e^{-.438445t} - .4850707\ e^{-4.561555t}$$

Notice that at $t = 0$, $p_{0,0}^{(0)}(0) = 1$ and $p_{0,1}^{(0)}(0) = 0$, as they should be. Furthermore, as $t \to \infty$ both expressions approach 0, which is in consonance with our knowledge that the two states $\underline{0}$ and $\underline{1}$ are not absorbing states and that, sooner or later, the system must visit state $\underline{2}$. The reader should also recognize that the above two equations give the time variation of the probability that the system, having started in state $\underline{0}$ at time 0, will be in state $\underline{i}$, $\underline{i}$ = 0 or 1, at time t <u>and</u> has never visited state 2. (This is the meaning of the (0) superscript on the p's.)

It is interesting to remark that both of the above expressions could have been obtained by a completely different reasoning through 'counting' the number of times the system enters state P_2. To this end introduce the 'counting' variable r along the arc P_1P_2, as shown in Fig. 3-9. It is easy to deduce that

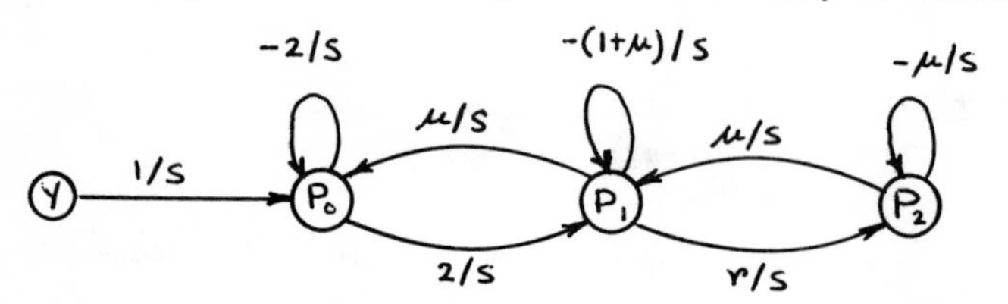

Counting the Number of Total System Failures

Figure 3-9

$$P_{0,0}(s,r) = \frac{s^2 + s(1+2\mu) + \mu(1+\mu) - r\mu}{A - r\mu(s+2)}$$

$$= \frac{s^2 + s(1+2\mu) + \mu(1+\mu)}{A\,[1-r\mu(s+2)/A]} - \frac{\mu r}{A\,[1-r\mu(s+2)/A]}$$

$$= \frac{s^2 + s(1+2\mu) + \mu(1+\mu)}{A}\left[1 + \frac{\mu(s+2)}{A}\,r + \left(\frac{\mu(s+2)}{A}\right)^2 r^2 + \ldots\right]$$

$$- \frac{\mu}{A}\left[r + \frac{\mu(s+2)}{A}\,r^2 + \left(\frac{\mu(s+2)}{A}\right)^2 r^3 + \ldots\right] \tag{3-6}$$

where $A = s^3 + s^2(3+2\mu) + s(2+3\mu+\mu^2) + 2\mu$.

The "zeroth" term of this series expansion is

$$P_{0,0}^{(0)}(s) = \frac{s^2 + s(1+2\mu) + \mu(1+\mu)}{A}$$

$$= \frac{s + (1+\mu)}{s^2 + s(3+\mu) + 2}$$

as before, since both numerator and denominator of the first expression are divisible by $(s+\mu) \neq 0$. A similar derivation would lead to $P_{0,1}^{(0)}(s)$.

Elapsed Times

Of major interest in the analysis of such queuing systems is the distribution of the elapsed time to the realization of certain events. For instance, what is the distribution of the elapsed time to the first complete system failure (in our example this means failure of both machines, i.e., being in state P_2)? This information may be of interest to management since it may indicate the duration of a 'working' system.

From Fig. 3-9 we deduce that

$$P_{0,2}(s,r) = \frac{2r}{A - r\,\mu(s+2)}$$

$$= \frac{2r}{A}\left[1 + \frac{\mu(s+2)}{A}\,r + \left(\frac{\mu(s+2)}{A}\right)^2 r^2 + \ldots\right].$$

where, as before, $A = s^3 + s^2(3+2\mu) + s(2+3\mu+\mu^2) + 2\mu$. Hence, the coefficient of r is

$$P_{0,2}^{(1)}(s) = \frac{2}{A} = \frac{2}{(s+\mu)(s^2+s(3+\mu) + 2)} = \frac{c_0}{s+\mu} + \frac{c_1}{s-s_1} + \frac{c_2}{s-s_2}$$

where s_1 and s_2 are the roots of the quadratic term, and are the same as given by Eq. (3-5); and

$$c_0 = \frac{2}{2 - 3\mu} , \quad c_1 = \frac{2}{(s_1 - s_2)(s \;\; + \mu)} , \quad c_2 = \frac{-2}{(s_1 - s_2)(s_2 + \mu)} .$$

At the particular value $\mu = 2$, the inverse Laplace transformation yields

$$p_{0,2}^{(1)}(t) = -.5\ e^{-2t} + .310633\ e^{-.438445t} + .189367\ e^{-4.561555t} \tag{3-7}$$

This is the time variation of $p_{0,2}^{(1)}$, which itself is a compound event: it denotes the probability that the system is in state <u>2</u> and has failed only once. The <u>conditional</u> time variation of the probability that it has failed only once is simply given by dividing $p_{0,2}^{(1)}(t)$ of Eq. (3-7) by the probability that the system is in state <u>2</u> which is given by $p_{0,2}(t)$ of Eq. (3-4),

$$(p^{(1)}(t) \mid \text{system is inoperative}) = p_{0,2}^{(1)}(t)/p_{0,2}(t)$$

$$= \frac{-.5\ e^{-2t} + .310633\ e^{-.438445t} + .18936573\ e^{-4.561555t}}{.2 - .3333\ e^{-2t} + .1333\ e^{-5t}} . \tag{3-8}$$

Although the expression is somewhat unwieldy it can be integrated numerically with relative ease. Upon normalization (through dividing by the total area under the curve, which is finite) we obtain the probability density function, from which the average duration to first failure is evaluated.

Naturally, if the system is inoperative the duration of its total failure is given by the holding time at that state, which is simply $h(\tau) = \mu e^{-\mu\tau}$.

The reader will remark that in approaching the problem of 'duration of failure' <u>via</u> the second reasoning (of Fig. 3-9), we have also developed a <u>counting</u> procedure for total failure. For instance $P_{0,0}(s,r)$ of Eq. (3-6) is already expressed as a series in r, say

$$P_{0,0}(s,r) = \sum_{k=1}^{\infty} b_k(s)\ r^k \ ;$$

in which $b_k(s)$ is the Laplace transform of the time variation of the probability of moving from state P_o to state P_o after having 'visited' node P_2 exactly k times (compare with the term $a_k(z)$ in the expansion of $t_{3,1}(z,r)$; p.). The normalizing factor is the probability of moving from P_o to P_o, given by $P_{0,0}(s,r)\Big|_{r=1}$, which was previously obtained as $P_{0,0}(s)$.

Analog Computation

A statistician reading the above analysis may remark that ingenious as SFGs may be for portraying the interaction among the various parts of the system, they have not yet produced anything new to him except perhaps the ease of visualizing the total system. We now point out that having the SFG representation is tantamount to having the wiring diagram for analog computation. Undeniably, this is a valuable 'bit of knowledge' from a practical point of view. Its value is equivalent to the statement that expressing the dynamics of a system in the form of a particular diagrammatic fashion is tantamount to having the computer program for digital computation (if the latter is true)!

Now, analog computers have traditionally been an engineering tool, in the same sense that the slide-rule has always been associated with engineers. There is no necessity for such restriction, and the theory of finite queues is an excellent example in which a statistician may be introduced to an engineering tool to the advantage of both.

An analog computer is composed principally of integrators, sign inverters, 'pots' (for potentiometers), wires, a source of D. C. current, and other pieces

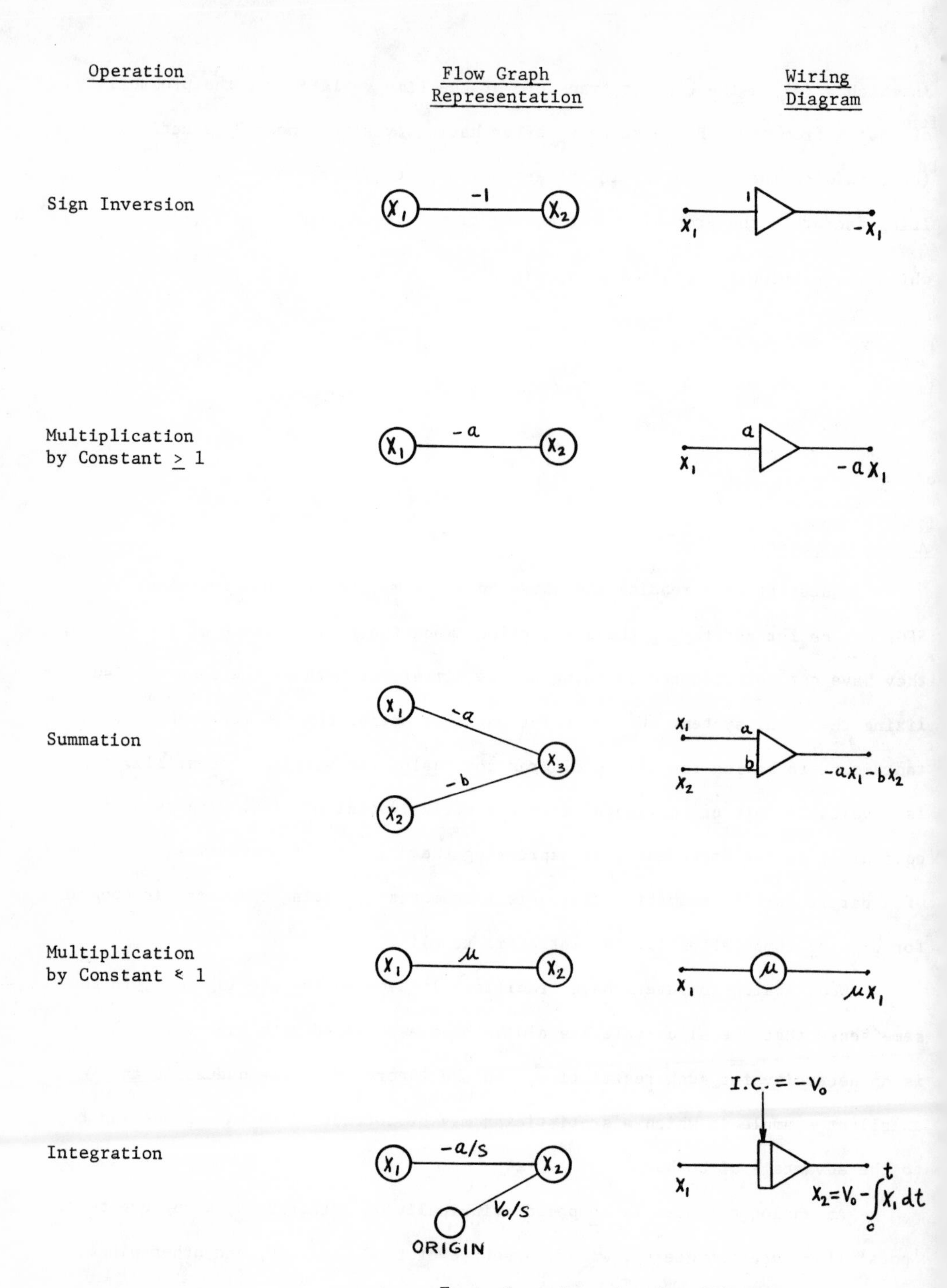

Equivalent Notation for SFG and

Wiring Diagram for Analog Computation

Figure 3-10

of equipment which do not concern us here (such as random noise generators, function generators, multipliers, etc.). The graphic symbols for analog computation, together with their equivalent SFG notation, is shown in Fig. 3-10.

We know, from our knowledge of Laplace transformation theory, that the operator 1/s indicates integration in the time domain (i.e., the inverse Laplace transform of $\frac{1}{s}$ G(s) is $\int_0^t g(t)\,dt$ where G(s) is the Laplace transform of g(t)). Therefore, it is not difficult to take one look at the SFG of Fig. 3-7 and immediately draw the wiring diagram of Fig. 3-11.

Of course, Fig. 3-11 is a network. It is the diagrammatic representation of the physical network composed of the various components of the analog computer for the simulation, in time, of the probabilities of our queuing problem. The reader has just been introduced to a still different kind of network!

A Re-Interpretation

The queuing problem which has served as our vehicle of illustration thus far can also be interpreted as a semi-Markov process, and a _different_ SFG can be drawn to represent the new relationships. We shall come back to this interesting way of looking at the problem after we have introduced semi-Markov processes in the next section.

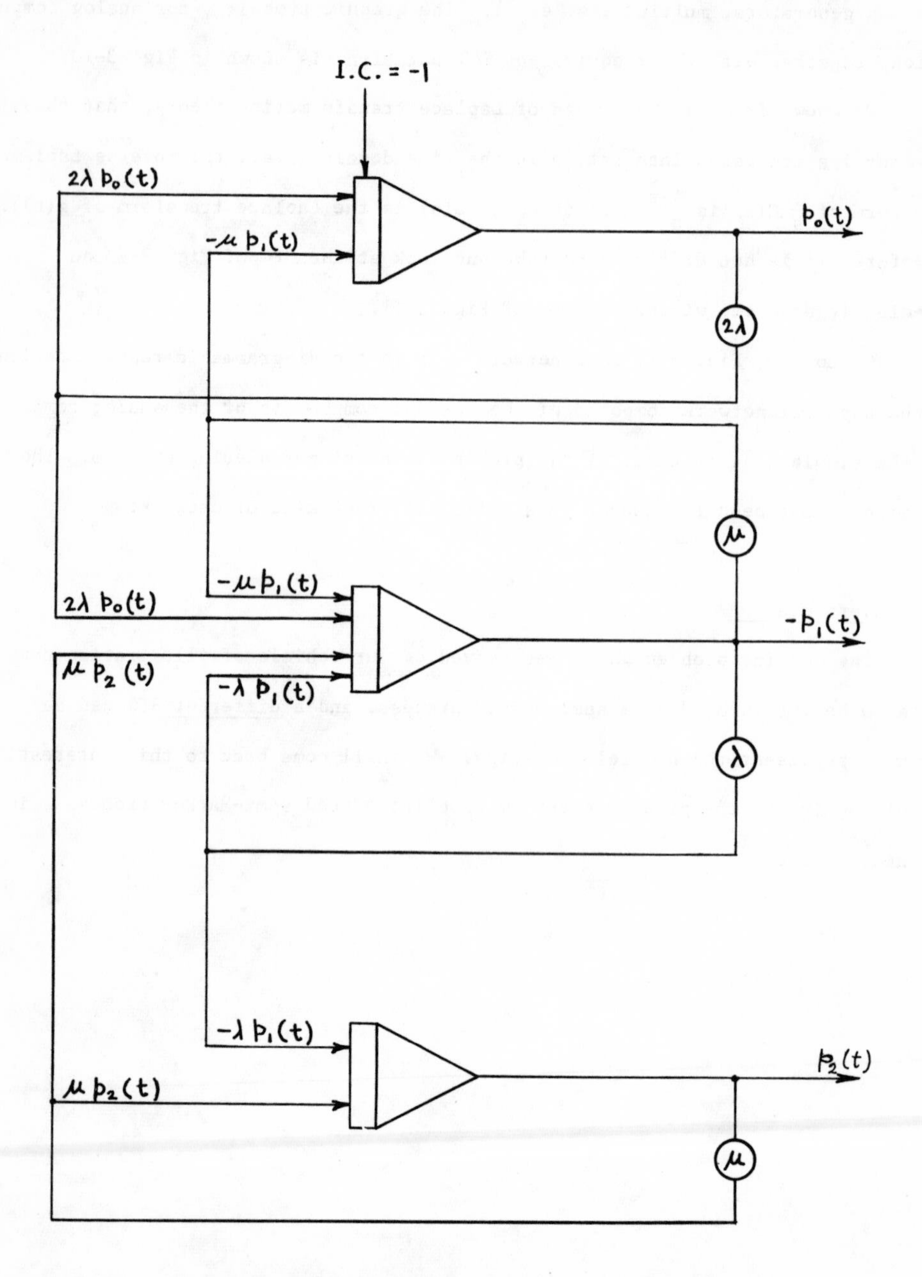

Realization of the SFG of Fig. 7

on an Analog Computer

Figure 3-11

REFERENCES TO CHAPTER 3

[1] Cox, D. R. and W. L. Smith, "Queues", Methuen's & Co., London, 1961.

[2] Flagle, C. D., et al, "Operations Research and Systems Engineering", Chapters 21 and 22, Johns Hopkins Press, 1960.

[3] Gue, R. L., "Signal Flow Graphs and Analog Computation in the Analysis of Finite Queues", Opns. Res., Vol. 14, No. 2, April 1966.

[4] Happ, W. W.,"An Application of Flow Graphs to the Solution of Reliability Problems", Physics of Failure in Electronics, Vol. II, Spartan Books, Inc., 1964.

[5] Howard, R., "Dynamic Programming and Markov Processes", MIT Press, 1959.

[6] Lorens, C. S., "Flowgraphs", McGraw-Hill, 1964.

[7] Mason, S. J., "Feedback Theory: Some Properties of Signal Flow Graphs", Proc. IRE, Vol. 41, No. 9, September 1953.

[8] ___________, "Feedback Theory: Further Properties of Signal Flow Graphs", Proc. IRE, Vol. 44, No. 7, July 1956.

CHAPTER 4

ACTIVITY NETWORKS AND THEIR GENERALIZATIONS†

Contents

§4.1 THE BASIC MODELS: "CPM" and "PERT"

Introduction

Traditionally, an activity network is a representation of two particular aspects of a project: (i) the precedence relationship among the activities, and (ii) the duration of each activity. Precedence, denoted by "$\prec$", is a binary relation which is transitive (i.e., if $u \prec v$ and $v \prec w$ then $u \prec w$), non-reflexive (i.e., $u \not\prec u$) and non-symmetric (i.e., if $u \prec v$ then $v \not\prec u$). It comes about from technological and other considerations; for example, one cannot build the roof of a house without first having built the walls, which in turn must be preceded by the foundation, etc. In the network representation, the nodes are usually taken to depict 'events', which are considered as well-defined occurrences in time (such as 'shipment received', or 'test completed', etc.); and the arcs to depict activities (such as 'construct foundation', or 'test for quality', etc.), though the roles of these two components can certainly be interchanged, in which case nodes would represent activities and arcs represent events. An activity usually consumes something such as materials, energy, skills, money, etc., and it usually takes time to accomplish, though we shall talk of 'dummy' (or pseudo) activities which consume nothing and are of duration zero.

---

† This chapter has greatly benefited from the very valuable comments of Professor A. Alan B. Pritsker, to whom I am grateful.

Furthermore, an arc leading from one node to another is directed (i.e., it is an arrow), and represents the activity that must take place after the realization of the node at the tail of the arrow in order for the node at the head of the arrow to be realized. Thus the direction of the arrow determines the precedence relation between the two nodes. On the other hand, precedence between two activities is represented by having the terminal node of the preceding activity be the initial node of the succeeding activity.

By virtue of the properties of the precedence relationship the resulting network is a directed, acyclic network, in which it is always possible to number the nodes such that an arc always leads from a small numbered node to a larger one. By construction, possibly through the use of dummy activities, the network has only one origin (node 1) and only one terminal (node n). Each node must have at least one arc leading into it and one arc going out of it except the origin (which has only arcs going out of it) and the terminal (which has only arcs leading into it). For purposes of machine computation, any two nodes may be connected by, at most, one arc. This requirement can be satisfied in one of two ways: either combine the two or more activities in parallel into one gross activity[†] or, better still, use dummy activities of zero duration and utility and dummy events, as shown in Fig. 4-1. Finally, since the precedence relationship is transitive, the realization of any node i necessitates the realization of all the preceding events and activities to that node.

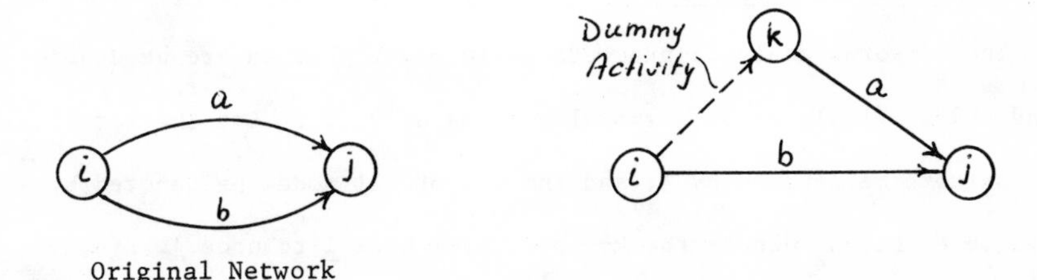

Introducing Dummy Activities for Parallel Paths

Figure 4-1

[†] This may not be desirable if it is preferred to keep the identity of the two activities separate, such as when each activity requires a different skill of labor.

In its simplest form, an activity network is better known under the acronym CPM (for Critical Path Method), in which the duration of each activity is a fixed constant. Analysis of such networks has been extensive, being, as they are, the simplest model that takes precedence into account. The study of their temporal characteristics reveals information on the minimum total duration of the project, the so-called 'duration of the critical path (CP)'; the earliest realization time of any node i, $t_i(E)$; (of course, the earliest realization time of node n, the last node, is synonymous with the earliest completion time of the whole project); the latest realization time of any node i; $t_i(L)$; and the resultant concepts of slack (or float) of an event i or of an activity (i,j).

In the few years that have passed since the CPM model was proposed, studies have moved beyond the stage of simple analysis to the stage of synthesis. It is possible, for instance, to determine the optimal duration of each activity to minimize a cost expression and achieve a target completion date for the whole project assuming, as it is reasonable to do, that there is a relationship between the duration of an activity and its cost.

On the other hand, the model was extended to take into account the problems of scarce resources. Such studies are mainly concerned with the determination of the optimal utilization of one or more scarce resource and still achieve desired completion dates for selected activities.

Temporal Analysis of Deterministic Activity Networks

The temporal analysis proceeds as follows. Let an arc be denoted by either its end nodes, (i,j), or by a symbol such as u, v,..., Let the set of all arcs of the network be denoted by A; and the set of all nodes be denoted by $N \triangleq \{1,2,\ldots,n\}$. Let Π_k denote the kth path from node 1 to node j. Finally, let the duration of activity (i,j) be a given constant y_{ij}. Now, by virtue of the precedence relationship, it is evident that the earliest realization time of node j is given by

$$t_j(E) = \max_k T(\Pi_k) \; ; \qquad t_1(E) \equiv 0 \; ; \quad j = 2,3,\ldots,n. \tag{4-1}$$

where $T(\Pi_k)$ is the 'length' of path Π_k and is equal to $\sum_{u \in \Pi_k} y_u$. Equation (4-1) is a definitional equation and simply requires the determination of the longest path from the origin, node 1, to each node $j \in N$ in a directed, acyclic network. But this is a problem which has been solved before (see § 1.3) in the discussion of the Shortest Path Problem. As we have seen before, it is better to apply recursively the equation

$$t_j(E) = \max_{i \in B(j)} \{t_i(E) + y_{ij}\}; \quad t_1(E) \equiv 0; \; j = 2,3,\ldots,n. \tag{4-2}$$

where maximization is taken over all the nodes that connect into j ; i.e., are in the set before j, $B(j)$.

As indicated before, the latest time a node must be realized in order to avoid a delay in node n is denoted by $t_i(L)$. We specify that:

either $t_n(L) = t_n(E)$, and then $t_1(L) = t_1(E) \equiv 0$;

or $\quad t_n(L) = t_n(s) > t_n(E)$, and then $t_1(L) > t_1(E) \equiv 0$;

where $t_n(s)$ denotes a specified completion time of node n. Without loss of generality, we may assume that $t_n(s) = t_n(L) = t_n(E)$. The other case in which $t_n(s) > t_n(L)$ can be easily deduced from this analysis by the simple addition of a constant equal to the difference $t_n(s) - t_n(L) \geq 0$.

By similar reasoning to above, and moving backwards from node n to node 1, we define $t_i(L)$ by

$$t_i(L) = \min_k \{t_n(L) - T(\hat{\Pi}_k)\} \; , \; t_n(L) = t_n(E); \tag{4-3}$$

where $\hat{\Pi}_k$ is the kth path from node n to node i (i.e., with all arrows of the network reversed), and $T(\hat{\Pi}_k)$ is its duration. We evaluate $t_i(L)$ recursively from

$$t_i(L) = \min_{j \varepsilon A(i)} \{t_j(L) - y_{ij}\}; \quad t_n(L) = t_n(E); \quad i = n-1,\ldots,1 \tag{4-4}$$

where the minimization is taken over all nodes j which are in the set of nodes occurring after _i_, $A(i)$. The difference

$$S_i = t_i(L) - t_i(E) \geq 0 \; ; \qquad i \; \varepsilon \; N \tag{4-5}$$

is the _slack time of event i_. It represents the possible delay in the realization of _i_ that causes no delay in $t_n(E)$.

Starting from the origin _1_ and moving _forward_ in a stepwise fashion, following Eq.(4-2), we evaluate $t_i(E)$ for $i = 2,3,\ldots,n$. This specifies the earliest completion time of the terminal, $t_n(E)$. Now, assuming $t_n(L) = t_n(E)$, we can move _backwards_ towards the origin _1_, using Eq.(4-4), and evaluate $t_i(L)$ for $i = n-1, n-2,\ldots,1$.

In this forward and backward movement there must be _at least one path_ whose slack (i.e., the slack of every activity and event along the path) is equal to zero. The node slack is given in Eq.(4-5); but the activity slack is a more subtle concept that demands a closer look.

In fact, there are more than one activity slack (or "float", as it is sometimes called). This is an immediate consequence of the fact that an activity possesses two end nodes, each one of which has two time values: an earliest realization time and a latest realization time. Hence, one can define _four_ activity slack values:

1. $S_{ij}(\text{Total}) = t_j(L) - t_i(E) - y_{ij}$; assumes that _i_ is realized early and that _j_ can be delayed as much as possible.

2. $S_{ij}(\text{Safety}) = t_j(L) - t_i(L) - y_{ij}$; assumes that _i_ is realized late and that _j_ can also be delayed.

3. $S_{ij}(\text{Free}) = t_j(E) - t_i(E) - y_{ij}$; assumes that both _i_ and _j_ are realized early.

4. S_{ij}(Independent) = max (0; $t_j(E) - t_i(L) - y_{ij}$); measures the absolute freedom in delaying the activity without affecting any other activity.

The various slacks were christened by Thomas[16]. As usual, the name is less significant than the precise definition of the variable.

As an illustration of the above concepts, consider the network of Fig. 4.2. Each node bears a vector of three numbers ($t_i(E)$, $t_i(L)$, S_i). The activity slacks are given in the following table:

Activity	S_{ij}(Total)	S_{ij}(Safety)	S_{ij}(Free)	S_{ij}(Indep.)	CP
(1,2)	3	3	0	0	
(1,3)	4	4	0	0	
(1,4)	0	0	0	0	X
(2,3)	8	5	4	1	
(2,4)	3	0	3	0	
(3,5)	4	0	4	0	
(4,5)	0	0	0	0	X

Notice that the CP is defined by the activities whose slack is identically equal to zero for all four categories.

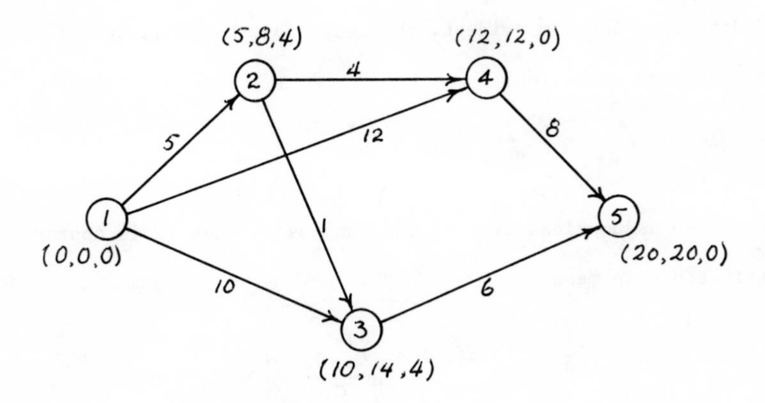

Figure 4-2

The PERT Model

The first generalization of the CPM basic network model involved the introduction of uncertainty in the duration of the activities. This new model has received wide acceptance in both government and industry and is better known as PERT (for Program Evaluation and Review Technique). Variations of PERT exist which usually involve minute and/or insignificant differences, mostly related to the manner of the summarization and presentation of the results of analysis, rather than to any fundamental differences in concepts. These variations are usually distinguished by new acronyms or new names, and the reader should be on his guard lest he confuse a new acronym with a new model.

When the durations of some of the arcs (activities) are random variables (r.v.'s), denoted by Y_u for arc u, the original approach proposed by the originators of PERT[11] ran as follows: let $\bar{y}_u = E(Y_u)$ be the expected value of the r.v.Y_u. Substitute everywhere the (so-called "certainty equivalent") expected value and determine the CP following the standard procedure of DAN's. Let π_C denote the CP(s). Assume that

(i) The activities are independent

(ii) The CP contains a "sufficiently large" number of activities so that we can invoke the Central Limit Theorem.[11]

Clearly, the length of the CP, $T(\pi_C)$, is identical with the time of realization of node n, denoted by T_n. Equally evident, the length of the CP is itself a r.v., being the sum of random variables,

$$T_n = \sum_{u \in \pi_C} Y_u = T(\pi_C).$$

The above two assumptions lead to the conclusion that T_n is approximately normally distributed with mean

$$\bar{T}_n = \sum_{u \in \pi_C} E(Y_u) = \sum_{u \in \pi_C} \bar{y}_u \quad ;$$

and variance

$$\sigma_n^2 = \sum_{u \varepsilon \pi_C} \sigma_u^2 \quad ;$$

where σ_u^2 is the variance of the r.v.Y_u.

The new concept introduced in the analysis of networks with probabilistic time estimates is that of probability statements (i.e., confidence statements) concerning the duration of the project. In particular, if $\Phi(x)$ denotes the standard normal d.f., i.e.,

$$\Phi(x) = \int_{\infty}^{x} (2\pi)^{-1/2} \, e^{-s^2/2} \, ds \quad ;$$

then the probability that event n will occur on or before a specified time $t_n(s)$ is given by

$$P_r\{T_n \leq t_n(s)\} = \Phi\{(t_n(s) - \bar{T}_n)/\sigma_n\} \quad ;$$

where $\bar{T}_n$ and σ_n are given above.

For the practical application of these ideas, the originators of PERT suggested that instead of asking for the d.f. of Y_u, for all u, or asking for "guesstimates" about the mean, $\bar{y}_u$, and the variance, σ_u^2, for all u, it is easier and possibly more meaningful to assume the following assumptions and simplifications in addition to the two assumptions made above:

(iii) The probability d.f. of Y_u can be approximated by the Beta distribution; i.e.,

$$dF_u(y) = K(y-a)^{\alpha}(b-y)^{\beta} \quad ; \; a \leq y \leq b \quad ;$$

where a and b are "location" parameters and α and β are "shape parameters" and K is a normalizing constant; and

(iv) The mean of the Beta d.f. can be approximated by $\bar{y}_u \simeq (a_u+4m_u+b_u)/6$; and the variance by $\sigma_u^2 \simeq (b_u-a_u)^2/36$; where m_u is the mode of the d.f. of Y_u, (the so-called "most likely duration").

Acceptance of these assumptions and simplifications leads one to conclude that for the application of PERT one needs to obtain the best estimates of the three parameters a,m, and b, which are labeled respectively, the "optimistic", "most likely" and "pessimistic" estimates of the duration of the activity. These parameters determine the estimates $\bar{y}_u$ and σ_u^2 which, in turn, are used in the determination of the CP and the evaluation of the probability of project duration.

The original PERT model depended rather heavily on the above-mentioned assumptions and simplifications, as well as on the assumption that the three parameters a, m and b can be extracted from the knowledgeable people for each activity in the network. Unfortunately, the assumptions and most of the conclusions based on them are open to serious questions that cast a grave shadow on their validity.

To illustrate the PERT approach and its calculations, suppose that the network shown in Fig. 4-3 is the precedence representation of a (very modest) project.

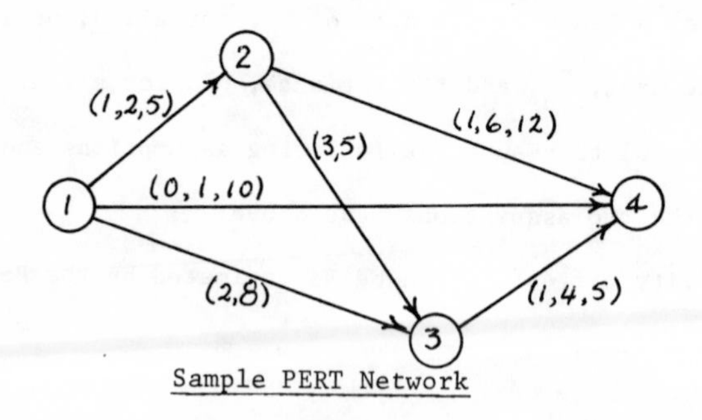

Sample PERT Network

Figure 4-3

The numbers on each arc are the possible durations of the activity, and are assumed equally probable.*

---

* Clearly, these numbers are different from the three required estimates a, m and b described above.

For example, activity (1,2) may consume 1 hour, 2 hours or 5 hours, each occurring with probability 1/3; similarly, activity (2,3) may consume 3 hours or 5 hours, each with probability 1/2; etc. Analysis proceeds in the following fashion.

Unquestionably, the total duration of the project is a random variable since it is dependent on the durations of the various activities, which are random variables themselves. For example, in Fig. 4-4, we present two possible realizations (from among a total of 324 different realizations), and it is seen that the duration of the CP is 8 in one case and 11 in the other. Consequently it is meaningless to speak of the total duration; rather, one should speak of some parameters (or statistics) of the distribution of that duration.

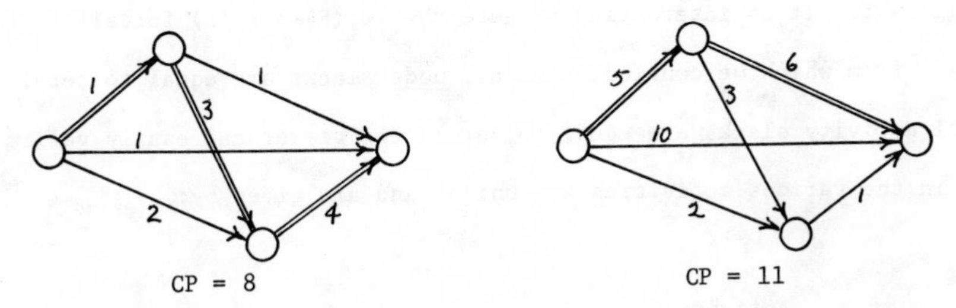

Two Possible Realizations of the PERT Network of Fig. 4-2

Figure 4-4

Thus, if one enumerates all possible realizations and evaluates the length of the CP in each realization, one can obtain the complete probability mass function of the duration (clearly, in this example it is a discrete probability function). This is an impossible task for any but the most trivial network and discrete probability mass functions for the arc durations.

The alternative to enumeration, if one still desires the complete probability distribution function, is Monte Carlo Sampling. In this case relatively few realizations are constructed based on a random sample from all possible realizations, and a histogram of frequency of durations is obtained to which a theoretical probability distribution function is fitted. Since the procedure is statistical all deductions are subject to error, which decreases as the size of the random sample increases.

If interest is mainly in the estimate of the mean duration of the project, various procedures are available which vary in their degree of accuracy of estimating the true mean. The simplest procedure (and the one proposed by the PERT originators) calls for substituting the mean duration of each activity for the random variable itself. In a sense, the mean is taken as the 'certainty equivalent' (or 'surrogate') of the random variable. Then analysis proceeds with these 'certainty equivalents' in a straightforward fashion, reminiscent of the analysis of CPM.

To illustrate, consider again the PERT network of Fig. 4-3. The expected value, $E(Y_u)$, for all six arcs, as well as the CP (the double-lined arcs) are shown in Fig. 4.5. It is interesting to note that $t_i(E) = t_i(L)$ for all $i = 1,2,3,4$; from which we conclude that all node slacks are equal to zero! And yet, not all activity slacks are equal to zero: the reader can easily verify that the slacks in the various activities are unique and are given by:

Activity (ij)	Slack S_{ij}
(1,2),(2,3),(3,4)	0
(1,3)	5/3
(1,4)	19/3
(2,4)	1.0

This example illustrates the subtle, and important difference between the concepts of node slack and activity slack.

The above statements concerning the slack in the activities are only pertinent if the activity times are constants or certainty is assumed, as, for example, in CPM networks; otherwise they are not true. In fact, the very estimate of the mean duration of 10 units is in error. An improved estimate (due to Fulkerson[8]) puts the value at 12.148; and a still better estimate (due to Elmaghraby[4]) increases it to 12.310. Unfortunately, all three estimators mentioned above are biased and it can be shown that they all underestimate the true mean. The correct value of the mean (obtained through complete enumeration of all 324 possible realizations!) is 12.314. Needless to say, the distribution of the duration of the project is given by the distribution of the random variable T_n, the earliest time of realization of node $\underline{n}$, where

$$T_n = \max_k T(\Pi_k)$$

and (4-6)

$$T(\Pi_k) = \sum_{u \in \Pi_k} Y_u ;$$

Π_k being the $k^{\underline{th}}$ path from $\underline{1}$ to $\underline{n}$. Note that the $T(\Pi_k)$'s are generally not independent since they will contain the same activities. Hence T_n is distributed as the maximum of a finite number of (usually dependent) random variables. [2]

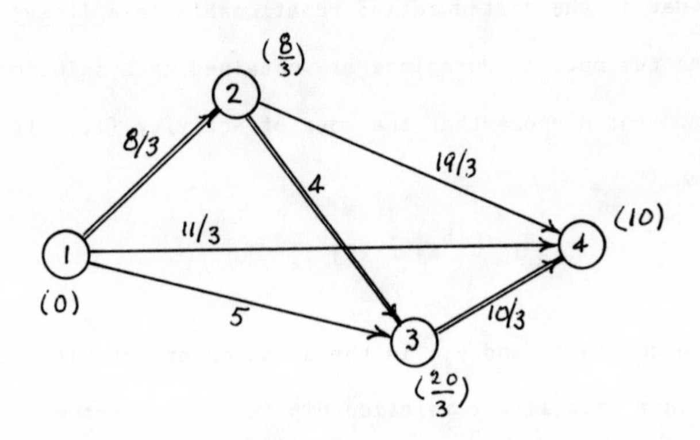

The PERT Network of Fig. 4-3 with $E(Y_u)$ Substituted

Figure 4-5

§4.2 A PROBLEM OF SYNTHESIS

We have alluded in our discussion of CPM above to the following question: how much time should be allowed each activity and, consequently, the whole project given a specified amount of funds, z^o; or, conversely, what is the minimal cost of a project that must be completed at a specified time λ? Such a question necessarily presumes the existence of different possible durations for some, if not all, of the activities in the project. As was mentioned before, such an assumption is not unrealistic. The majority of the activities in the industrial, government and military projects can be accomplished in shorter or longer durations by increasing or decreasing the resources (such as facilities, manpower, money, etc.) available to them.

It turns out that if the cost-duration relationship is a linear relationship for each activity then the optimal durations are obtained as a solution to a very simple linear program. For suppose that the cost of activity (i,j) is given by

$$c_{ij} = b_{ij} - a_{ij}y_{ij}$$

where a_{ij}, $b_{ij} \geq 0$ are constants and y_{ij} is the duration of activity (i,j). Furthermore, suppose that this linear relationship is valid for the range

$$d_{ij} \leq y_{ij} \leq D_{ij}. \tag{4-7}$$

Finally, let t_i denote the earliest time of realization of node i, and A the set of all arcs of the network, then, clearly,

$$t_j \geq t_i + y_{ij},\ i \rightarrow j, \text{ for all arcs } (i,j)\varepsilon A \tag{4-8}$$

$$t_1 = 0,\ t_n = \lambda,$$

where λ is a specified constant.

The Ineq. (4-8) simply states that if event j occurs after event i, and an activity (i,j) exists in the network, then the difference between the two times of realization must be at least equal to the duration y_{ij} of that activity. Putting $t_n = \lambda$ merely specifies the desired duration of the project. Naturally, we wish to minimize

$$\sum_{(i,j)\varepsilon A} c_{ij} = \text{constant} - \sum_{(i,j)\varepsilon A} a_{ij} y_{ij} \qquad (4\text{-}9)$$

Equations (4-7) to (4-9) specify a linear program in the unknowns (t_i) and (y_{ij}). If λ is permitted to vary[†], we have a parametric LP.

Interestingly enough the expeditious solution of this LP is achieved through a flow-network interpretation! Such networks were discussed in §2. above. The formulation is an ingenious interpretation of costs as capacities, and is due to Fulkerson[7]. It is a special case of the Optimal Circulation problem discussed in greater detail in §2.5 above.

There is no equivalent in PERT, with its probabilistic time estimates of activity durations, to the synthesis problem of CPM. The question has no content in the general probabilistic case, but may possess meaning in special applications such as when all the activities possess the same p.d.f., because then one may ask for the optimal value of a parameter (or parameters) of that p.d.f. This question is still open to research.

§4.3 GENERALIZED ACTIVITY NETWORKS

The conceptual models of the PERT-CPM variety are severely restricted from a logical point of view and in many instances fall short of adequately representing a wide variety of projects, especially of the research and development kind. For example, there is an implicit determinateness in the existence of all events

† It is easily seen that $(t_n | y_{ij} = d_{ij}) \leq \lambda \leq (t_n | y_{ij} = D_{ij})$, and that the LP is infeasible for any λ outside this range.

and activities - a determinateness which is exemplified by the calculations of the CP. To see this, consider Fig. 4-3 again. Activities (2,3) and (2,4) _must_ be undertaken, sooner or later, after the realization of event _2_. Uncertainty in PERT is not related to the eventual undertaking of these activities; only to their durations. On the other hand, both activities (1,3) and (2,3) _must_ be undertaken before event _3_ obtains. In fact, activity (3,4) cannot take place except after the completion of _both_ activities (1,3) and (2,3). The terminal node _4_ (or, in general, the terminal node _n_ of a PERT-CPM network) occurs _after all project activities and events have taken place_. Needless to say, industrial and economic systems are replete with different logical classes of events, uncertain activities, etc., which cannot be handled by the above-mentioned models.

For instance, suppose that node _1_ of Fig. 4-2 represents the event: submit price quotations on Projects A and B. Furthermore, suppose that there is a reasonable chance of winning a contract on either project A or project B, _but not on both_. How can we represent such eventualities? Obviously, the PERT-CPM models are inadequate unless we are willing to construct an activity network for each possible turn of events - a procedure which is at best clumsy and at worst computationally infeasible.

As another illustration, consider a project in which, upon reaching a certain stage of experimentation, progress is either continued along previously laid-out plans or a fundamental flaw in the product is discovered and the engineers are "back to their drawing boards". Here is an instance of _feedback_, previously banned from the PERT-CPM models.

In fact, it is not difficult to find situations in which the logical structure of PERT-CPM is highly inadequate. Examples abound in the areas of computer programming systems, bidding and contracting situations, missile count-down procedures, investment selection, production systems, and many others.

In an effort to cope with the need for such expanded representations, 'Generalized Activity Networks' (GAN) were introduced by Elmaghraby[3] in 1964 and were further extended in the subsequent years by Pritsker and Happ[12], Whitehouse[17], and Elmaghraby[5]. The basic formulation of the model is as follows.

There are two kinds of nodes: ordinary nodes and 'status' nodes; we shall always refer to the former as 'nodes' and to the latter as 'status nodes'. Each node may be considered a 'receiver' when it is at the head of the arrow and a 'source' or 'emitter' when at the tail of the arrow. There are three kinds of receiving nodes (see Fig. 4-6):

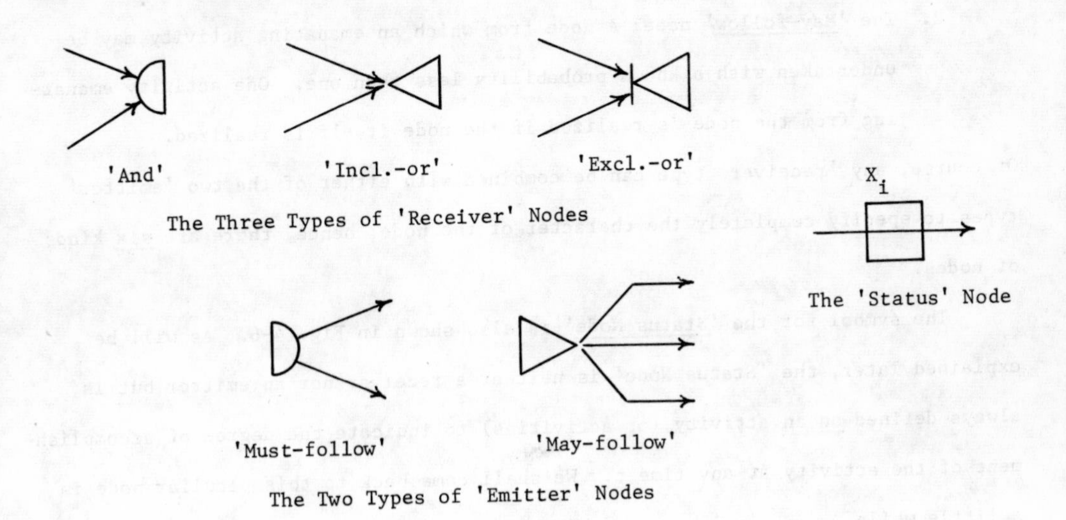

Types of Nodes

Figure 4-6

1. The Logical 'And' node: The node (i.e., event) will be realized when all arcs leading into it are realized.
2. The logical 'Inclusive-or' node: The node will be realized if any one arc or combination of arcs leading into it are realized. The node will be realized at the earliest completion time of the activities leading into it.

3. The logical 'Exclusive-or' node: The node will be realized if one and only one of the arrows leading into it is realized at any given time.

All nodes (except terminal nodes) will have arcs emanating from them, and two types of 'source' or 'emitter' nodes are defined depending on whether all the activities must be undertaken subsequent to the realization of the node, or the occurrence of some of these activities is probabilistic in nature (see Fig. 4-6).

1. The 'Must-follow' node: A node from which all emanating activities must be undertaken.
2. The 'May-follow' node: A node from which an emanating activity may be undertaken with a known probability less than one. One activity emanating from the node is realized if the node itself is realized.

Of course, any 'receiver' type can be combined with either of the two 'emitter' types to specify completely the character of the node; hence, there are six kinds of nodes.

The symbol for the 'Status Node' is also shown in Fig. 4-6. As will be explained later, the 'Status Node' is neither a receiver nor an emitter but is always defined on an activity (or activities) to indicate the degree of accomplishment of the activity at any time t. We shall come back to this peculiar node in a little while.

The generic element of GAN is shown in Fig. 4-7. It is composed of two nodes

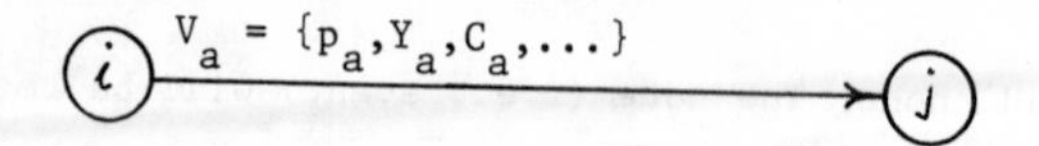

The Generic Element of GAN

Figure 4-7

i and j and the directed arc u = (i,j) joining them. On the arc is defined a vector of parameters, V_u, of dimensionality $m \geq 2$, of which we have shown only three in Fig. 4-7:

p_u: The probability that arc u (i.e., activity a) will be realized (which need not be synonymous with the probability of realization of node j since there may be other paths to j) given that node i is realized.

Y_u: A random variable representing the duration of arc u, assuming u is realized. Y_u is assumed having the probability density function $h_u(\tau)$.

c_u: A cost function which may, or may not, be dependent on the duration Y_u.

The first two parameters are always specified; other parameters (such as cost, consumption of one or several skills of labor, consumption of machine capacity, etc.) may be specified. If the activity occurs with certainty, the p_u may be dropped from explicit notation, though it is always implicitly recognized. If all arcs occur with certainty the GAN model reduces to the more familiar PERT model. If, furthermore, all durations (Y_u) are degenerate of the form: $g_u(\tau_0) = 1$ for $Y_u = \tau_0$ and $g_u(\tau) = 0$ otherwise, i.e., if the duration of each activity is deterministic, the model further reduces to the more elementary CPM model.

For the sake of clarity of exposition of the remainder of this discussion we will assume that the vector V is of dimensionality 2, i.e., $V_u = (p_u, Y_u)$; in which Y_u is generic of all 'additive' parameters (i.e., the value for two arcs in series is equal to the sum of the two values). As we shall see below, multiplicative parameters may also be given. It is emphasized, however, that this is done only for convenience and that generalization to higher dimensional vectors is straightforward. In fact, a production example to be presented later will use three parameters: probability, time, and cost.

The analysis of GAN is dependent on the algebra by which arcs are combined together in 'equivalent' arcs, similar to what was previously done in the discussion of signal flow graphs. The basic algebra for GAN is summarized in Fig.4-8. The following remarks should be helpful in understanding this set of rules.

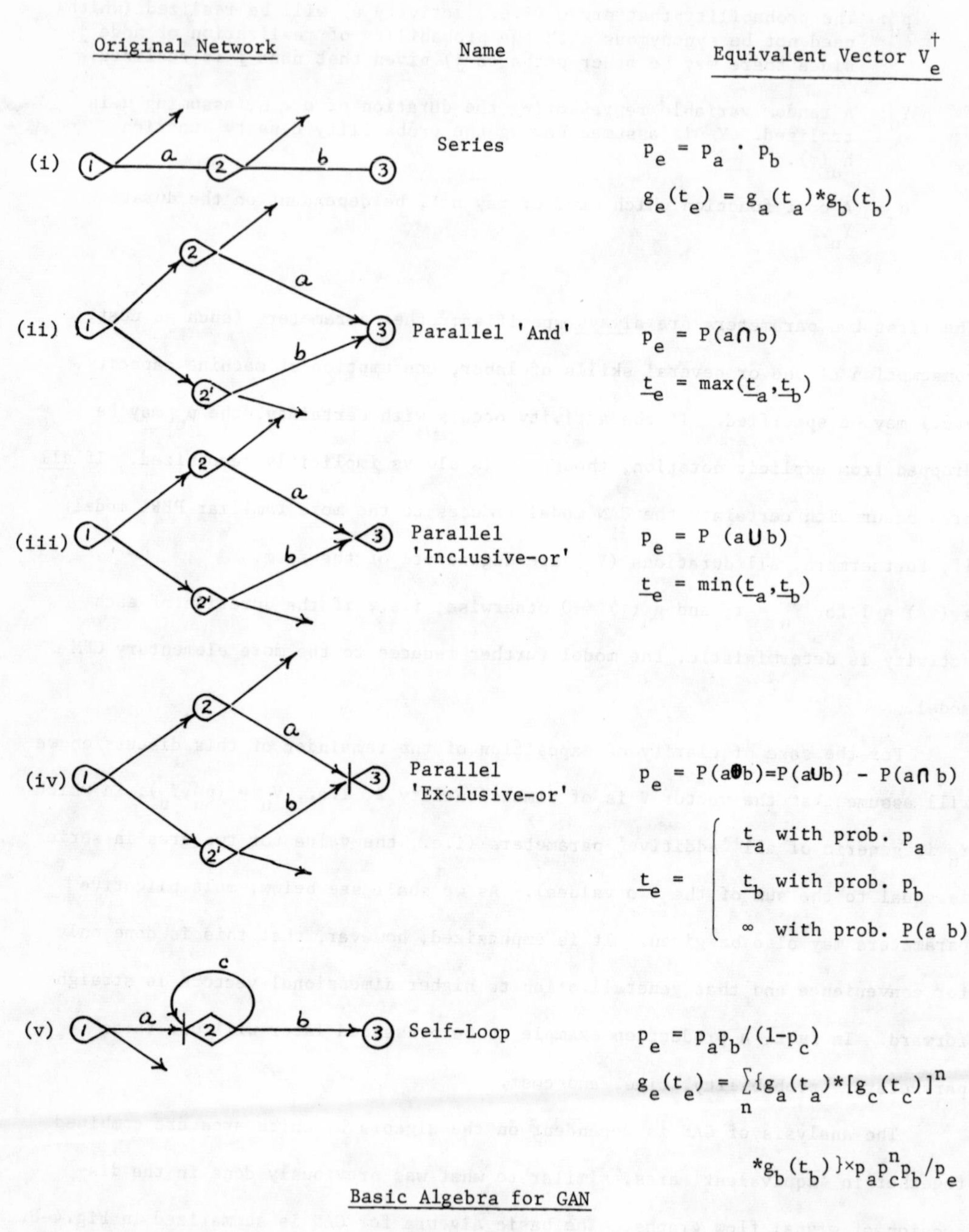

Original Network
Name
Equivalent Vector $V_e^{\dagger}$
(i)
a
b
1
2
3
Series
$p_e = p_a \cdot p_b$
$g_e(t_e) = g_a(t_a)*g_b(t_b)$
(ii)
2'
Parallel 'And'
$p_e = P(a \cap b)$
$\underline{t}_e = \max(\underline{t}_a, \underline{t}_b)$
(iii)
Parallel 'Inclusive-or'
$p_e = P(a \cup b)$
$\underline{t}_e = \min(\underline{t}_a, \underline{t}_b)$
(iv)
Parallel 'Exclusive-or'
$p_e = P(a \oplus b) = P(a \cup b) - P(a \cap b)$
$\underline{t}_e = \begin{cases} \underline{t}_a \text{ with prob. } p_a \\ \underline{t}_b \text{ with prob. } p_b \\ \infty \text{ with prob. } \underline{P}(a\ b) \end{cases}$
(v)
c
Self-Loop
$p_e = p_a p_b/(1-p_c)$
$g_e(t_e) = \sum_n \{g_a(t_a)*[g_c(t_c)]^n *g_b(t_b)\} \times p_a p_c^n p_b/p_e$

Basic Algebra for GAN

Figure 4-8

† The symbol * indicates convolution of the two probability functions, as in (i). Whenever the durations themselves are stated, rather than their d.f., it is understood that we refer to the random variables, hence they are written in boldface as in (ii) and (iii). The symbol $\oplus$ is the 'ring sum', defined as the occurrence of one, and only one, event.

1. By definition, each node (except terminal nodes) must have at least one arc (i.e., activity) emanating from it. In the case of more than one arc, it is possible that some arcs occur with certainty (hence their p_u's = 1) while others occur probabilistically (hence their p_u's are strictly > 0 and < 1). We adopt the convention that the sum of the probabilities of the activities whose p_u's are < 1 must add up to 1. Consequently, the sum of probabilities of the set of arcs emanating from any node must add up to an integer ≥ 1. Such requirement is neither illogical nor infeasible. On the contrary, the converse is true, since it is meaningless to say that after the realization of a node the system may realize one activity only with probability .25, say. The natural question is: what happens in the remaining 75% of the time? The above convention demands the answer to this question.

2. It is impossible to have a feedback loop on an 'And' type node because that would imply that an activity must be realized before it is realized! Furthermore, a feedback loop is meaningless on an 'Inclusive-or' node since the feedback arc will always be undertaken after a non-feedback arc, and hence after the node itself, has been realized. Consequently the node can be replaced by an 'Exclusive-or' node.

3. The realization of an 'And' node is dependent on the simultaneous realization of all the arcs leading into it, and the time of the equivalent arc is the maximum of a finite set of random variables. This is to be contrasted with the 'Inclusive-or' node whose realization is dependent on the realization of at least one of the arcs leading into it; and the time of the equivalent arc is the minimum of a finite number of random variables.

4. In GAN it is possible that some nodes are never realized. If arcs a, b, c, ... lead into a node, then:

an 'And' node may not be realized with probability $1 - P(a \cap b \cap c \cap \ldots)$

an 'Inclusive-or' node - ditto - with probability $1 - P(a \cup b \cup c \cup \ldots)$

an 'Exclusive-or' node - ditto - with probability $1 - P(a + b + c + \ldots)$

This should come as no surprise to the reader since it was precisely to accommodate such eventualities that GAN was developed in the first place.

5. Of all the newly defined nodes, by far the easiest to handle mathematically is the 'Exclusive-or' type node. In fact, the analysis of GAN is greatly facilitated when all the nodes of the network are of the 'Exclusive-or' type. The reason for such facility is that a simple transformation changes the network into a signal flow graph; hence, all what we have previously said about SFGs can be applied directly to the analysis of such networks. We shall discuss this in greater detail below.

But what is important to note at this time is that it is always possible to transform other nodes into 'Exclusive-or' nodes, at the expense of enlarging the network. For example, an 'And' node can be replaced by two 'Exclusive-or' nodes, one representing the eventuality that the 'And' node is realized, the other when it is not. Similarly, an 'Inclusive-or' node may be replaced by an 'Exclusive-or' node provided that the incoming activities are transformed into the various possible realizations of subsets of the original activities.

Therefore, it seems that the price paid for the use of an already established theory (that of SFG) is the enlargement of the original logic of the network. Whether such trade-off is advantageous or not can be answered only from empirical evidence.

As an example of the utility of GAN for the representation of complex logical relationships, consider a slightly modified investment problem given by Canada[1]: A firm has the choice between retaining an old plant at a cost of $10 (the cost is in millions of dollars) for major renovations, or scrapping the old plant and building a new one at a net cost of $35. The expected life of the renovated old plant is 3 years and of the new plant is 6 years. The future is uncertain: the market demand may be high or low (it is assumed that these are the only two alternatives), and the odds are even on it being one or the other. If the market is high the firm can realize $15 per year with the new plant but only $10 per year with the old plant; however, if the market is low, the firm can realize only $8 per year with the new plant and $6 with the old plant.

In addition to the above two alternative choices, the suggestion was made by the firm's sales vice president to enlist the help of some market research (MR) organization, at a cost of $1, to perform some studies concerning the future of the market, which, in turn, would help management to make the choice between the old and new plants. The study would take 3 months (.25 years). Of course, everybody realizes that the MR group is liable to commit an error, but the sales vice president had an 85% confidence in their judgment. That is, it was felt by him that if the MR organization predicts a 'high' market, the probability is .85 that the market will in fact be high, and is only .15 that it will be low; similarly if the prediction is for a 'low' market.

What is the best decision, assuming a discount factor α = .8 (which is equivalent to an 'internal rate of return' of 25% which, in turn, reflects the existence of other lucrative opportunities for investment)?

For the sake of economy we shall not give the detailed and complete answer to this question. However, the representation of this investment problem in GAN format is shown in Fig. 4-9. Notice that there are, in fact, two decision points. The first has been highlighted by labeling it as a 'decision' node, while the second occurs after the MR prediction has been made, and has not been highlighted in the GAN representation. There is a difference in time between these two decisions, but this difference has been ignored in the subsequent analysis. Hence, the node labeled 'old' represents the realization of 'retaining the old plant' through either a decision at the outset or after waiting for the MR prediction. Similarly for the node 'new'.

Several (but not all) node types are used in Fig. 4-9. The diagram is self-explanatory except at two points.

Consider what happens after the MR makes its forecast. Suppose it forecasts a "Hi" market (a similar reasoning applies to the case of forecasting a "Lo" market). At this point, management must decide on whether to keep the 'old' plant or construct a 'new' one; hence the probabilistic branching with probabilities q_2 and $1-q_2$. The simulation of the actual realization of a "Hi" or a

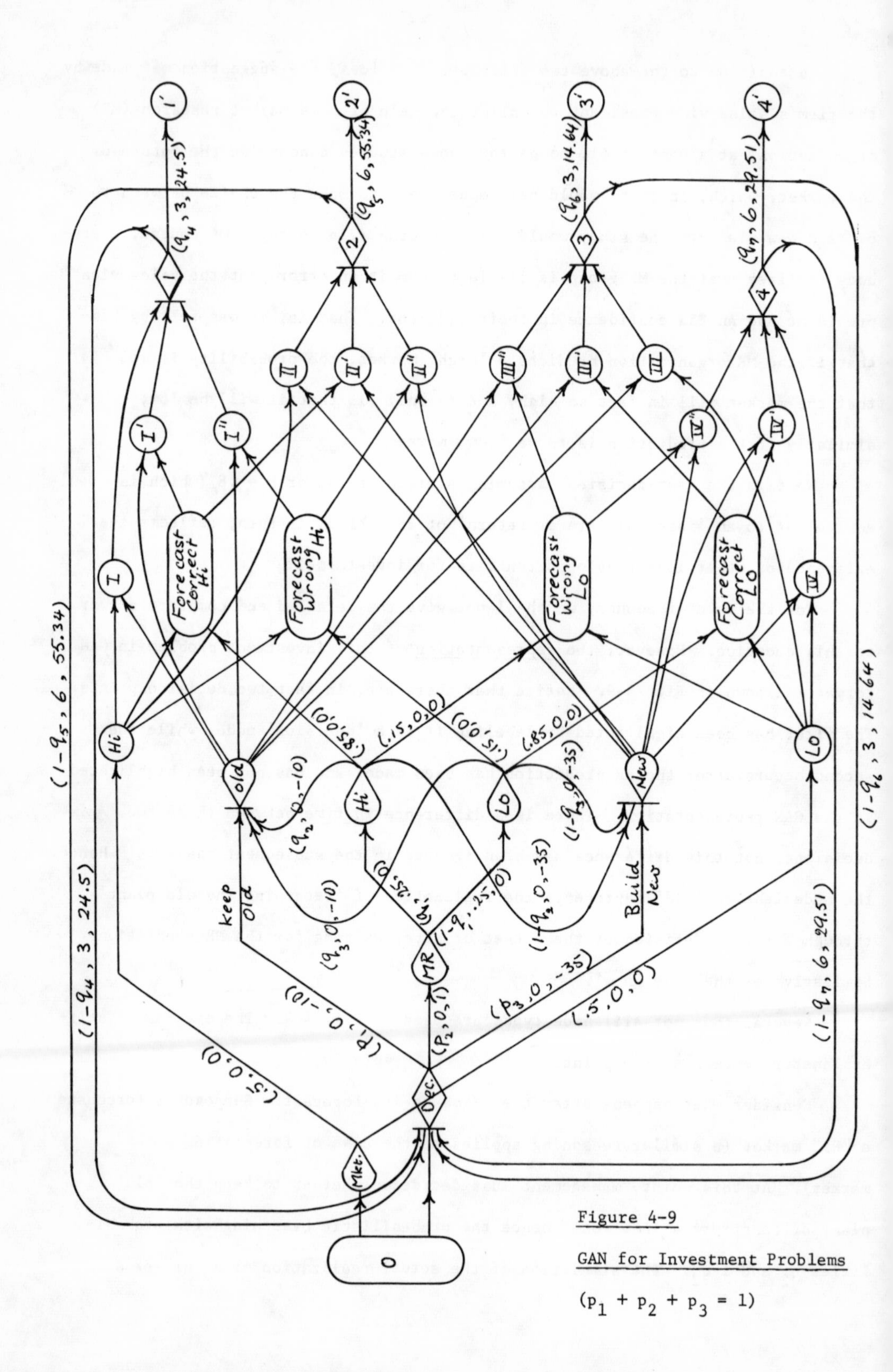

Figure 4-9

GAN for Investment Problems

$(p_1 + p_2 + p_3 = 1)$

"Lo" market is achieved by two 'And' type nodes labeled "Forecast Correct-Hi" and "Forecast Wrong-Lo".

Consider next node 1. It is realized if either node I, or node I' or node I" is realized, each one of which represents the combined realization of a "Hi" market together with the choice of the 'Old' plant. But each node represents this simultaneous realization via a different set of conditions: node I by the outright selection of the "Keep Old" alternative, node I' after the MR forecast turns out to be correct, and node I" after the MR forecast turns out to be wrong.

Now, assuming the realization of node 1, if the company decides to terminate the production of this particular item at the end of three years, the system will transfer to node 1', and under the conditions of 'old plant plus high demand' it will realize $\$10 + .8 \times \$10 + (.8)^2 \times 10 = \$24.40$. The probability that the company will 'quit this product' at the end of three years is q_4; and with probability $1 - q_4$ the company will 'continue in business'. Then it will return to its first position; hence the feedback loop. Of course, care has to be taken in combining the monetary benefits from this 'second time around' with the monetary benefits from the 'first round' because of the discount factor $\alpha = 0.8$.

The 'Status' Node and 'Conditional Parallel Progress'

The 'Status Node' (SN) is a device to extend the concept of GAN to the following interesting case: it often happens that a set of two or more activities can be started following the occurrence of an event, but that progress along some of these activities is conditional upon progress along one (or several) 'controlling' activity (or activities) in the set. If progress along the controlling activity is stopped, progress along the other activities of the set is also stopped. More often than not, all the activities which have been run concurrently, that is, in parallel, revert back to their 'start' conditions, and a new attempt is undertaken. This is the origin of the title 'Conditional Parallel Progress'. It is assumed that the cycle is repeated as often as progress along the controlling activity is stopped.

This pattern of behavior is not uncommon in research and development projects, in the testing of large systems, and in other undertakings whose outcome is uncertain. For example, the design of the transport vehicle for a high speed urban transportation system may proceed in parallel with the design of the road bed. At some point in time a check must be made on the latter activity (which is the controlling activity). If the design of the road bed does not satisfy certain minimal requirements, the whole concept of the transport vehicle will be in need of revision. Progress along the original design concept would be halted, and design along a different concept initiated.

The representation of conditional parallel progress utilizing the SN for the simple case of two activities in parallel is shown in Fig. 4-10. Originally,

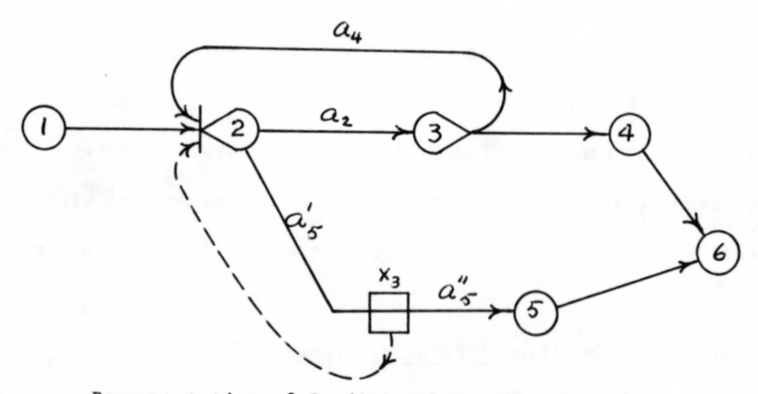

Representation of Conditional Parallel Progress

Figure 4-10

activities a_2 and a_5 could be run in parallel. However, activity a_2 (the controlling activity) terminates with a test (or an experiment, or any activity whose outcome is uncertain) which, if fails, returns both activities to their starting point, node 2.

To represent the dependence of activity a_5 on the outcome at node 3, the status node X_3 is introduced on arc a_5. The dotted line between SN X_3 and node 2 indicates the return to node 2 in case of failure at node 3 (hence the subscript 3 on the SN).

The conditional parallel progress represents a gamble to save time. Should the gamble fail, the cost of performing all the parallel activities would simply be wasted. If the gamble succeeds, however, a considerable saving in time would have been effected. For this reason, conditional parallel progress should not be contemplated, let alone undertaken, if such saving in time is not realizable, that is, if the duration of a_5 plus all subsequent activities dependent on its completion (activity a_6 in Fig. 4-10) is shorter than the duration of the activities subsequent to the controlling activity a_2.

Equally obvious, the undertaking, or otherwise, of activity a_5 in parallel with activity a_2 is not only dependent on the durations mentioned above, but also on the probability of failure at node 3. (If such probability varies with time in some regular fashion, say monotonically increasing or decreasing, it would be easy to determine the 'cut-off' probability beyond which it is (or it is not) uneconomical to undertake conditional activities in parallel).

The utility of extending GAN concepts to encompass the case of conditional parallel progress is now apparent: it assists the system's designer to consider, in an explicit fashion, the effect of one additional, and important, factor and to make a decision based on objective analysis.

In effect, the designer can decide at each realization of node 2 (or, in general, the node from which the controlling activity starts) whether or not to undertake the dependent activities in parallel or in series. We shall not pursue this analysis here since it is elementary in nature. The interested reader may consult ref. [5].

All 'Exclusive-or' GANS: GERT Networks

If all the nodes of a GAN are of the 'Exclusive-or' type on their receiving side (and we have previously indicated that all other nodes can be reduced to 'Exclusive-or' nodes, at least theoretically), then a GAN is in fact a SFG and the the activity network is a semi-Markov process. This fundamental observation is due to Pritsker and Happ[12], who gave the resulting network the identifying acronym "GERT", for Graphical Evaluation and Review Technique.

Naturally, GERT networks are valid representations of a much wider class of problems than mere activity networks with 'Exclusive-or' nodes, being, as they are, SFGs of semi-Markov processes. For other applications in reliability, queuing, decision-making, etc., the interested reader is referred to the growing literature on GERT, such as refs. [12] and [14]. On the other hand, the desire to represent and to be able to quantitatively evaluate more complex logical operations (such as switching functions, limited feedback loops, reset decisions, etc.) which are quite common in research and development endeavors, caused the GERT networks to move more recently from the realm of the simple 'Exclusive-or' node networks which are analytically analyzed to more complex logical networks which can be studied only through the technique of Monte Carlo simulation. A computer program, GERTS II, which is written in FORTRAN IV, is available, which is undergoing major changes at the time of this writing[13].

We do not propose to give here a detailed account of these developments [17]. Rather, the remainder of this section is devoted to establishing the relationship between GANs of this particular structure and semi-Markov processes (SMPs) on the one hand and the SFG representation of these processes on the other. In the process of establishing these relationships we shall derive several of the GERT characteristics in a different way.

First, we give the definition of an SMP, and it will become immediately obvious to the reader that all-'Exclusive-or' GANs are SMPs. Then we demonstrate how such processes can be easily represented as SFGs.

An SMP is a probabilistic system that makes its transitions from any state $\underline{i}$ to any other state $\underline{j}$ (which may be equal to $\underline{i}$) according to the transition probability matrix of a Markov process, but whose time between transitions could be an arbitrary random variable that depends on both $\underline{i}$ and $\underline{j}$.

The reader will recall from our discussion of the inventory problem (see section §3.1) that we then introduced a 'holding time' function $h_{ij}(\tau)$, denoting the probability (density or mass) function of the <u>duration</u> of the transfer from state $\underline{i}$ to state $\underline{j}$; $h_{ij}(\tau) \geq 0$, $\int_{\tau=0}^{\infty} h_{ij}(\tau) = 1$, which is permitted to depend on both $\underline{i}$ and $\underline{j}$. At that time, we remarked that in regular discrete time Markov processes $h_{ij}(\tau)$ is of the degenerate type since the time of transition is a constant equal to one time unit (by so defining the unit of time to be equal to the time of transition). In SMPs, this simple structure is no more the case: the duration is a random variable Y_{ij} which does not possess a degenerate d.f.

If it is of any help to the reader in visualizing an SMP, he may think of it in the following fashion. When the system reaches state $\underline{i}$ it immediately decides on its next state $\underline{j}$. This decision is made randomly according to the probabilities (p_{ij}) of the transition matrix. Now the system chooses to wait (i.e., 'hold') in state $\underline{i}$ for a time τ, where τ is picked randomly from the d.f. of Y_{ij}.

With these definitions and interpretations in mind, it is not difficult to see immediately that a GAN with all 'Exclusive-or' nodes is an SMP. Recall that the generic element of GAN (see Fig. 4-5) is an arc joining two nodes, $\underline{i}$ and $\underline{j}$, on which is defined a vector $V_{ij} = (p_{ij}, Y_{ij}, c_{ij}, \ldots)$. For any node $\underline{i}$, the p_{ij} is the probability that the system will transfer to node $\underline{j}$ (or equivalently, that (i,j) will be realized), with the stipulation that if $p_{ij} < 1$, then $(\sum_j p_{ij} | p_{ij} < 1) = 1$. This is precisely the definition of the transition probabilities in the stochastic matrix of the imbedded Markov process. Of course, the duration of the activity, Y_{ij}, plays the role of the 'holding time' in SMP. This is the reason for having denoted its p.d.f. by the same symbol $h_{ij}(\tau)$ in the previous discussion.

Let us next establish the representation of SMPs by SFGs. This discussion could have been included in the previous section on SFG theory. Its location is purely a matter of taste and style of exposition.

An observer watching the system in state $\underline{i}$ can only infer the unconditional p.d.f. of the waiting time at state $\underline{i}$, denoted by $w_i(t)$, where

$$w_i(t) = \sum_{j=1}^{n} p_{ij} h_{ij}(t) \qquad (t \geq 0)$$

Let $\bar{\tau}$ denote the unconditional mean waiting time at i, while $\bar{\tau}_{ij}$ denotes the conditional mean waiting time between i and j. Clearly

$$\bar{\tau}_{ij} = \int_o^{\infty} \tau h_{ij}(\tau)\, d\tau \text{ , and}$$

$$\bar{\tau}_i = \sum_{j=1}^{n} p_{ij}\bar{\tau}_{ij}$$

Let $W_i(t)$ denote the cumulative distribution function (cum. d.f.) of the unconditional waiting time at i, i.e.,

$$W_i(t) = \int_o^t w_i(\tau)\, d\tau \text{ ,}$$

and let

$$\tilde{W}_i(t) = 1 - W_i(t)$$

denote the complementary cumulative d.f. Finally, let $\phi_{ij}(t)$ denote the conditional probability that the system is in state j at time t given that it was in state i at time zero. We shall refer to $\phi_{ij}(t)$ as the 'interval transition probability'. Note the difference between ϕ_{ij} and p_{ij}: the latter is the conditional probability of a single transition, while the former is the conditional probability of being in j at time t, which may involve several transitions.

In fact, this latter probability can be stated in a recursion equation as follows: either the system started at j and never made a transition during t, and the probability of that is $\delta_{ij}\tilde{W}_i(t)$ (where δ_{ij} is Kroeneker delta), or it started at some state i; i = 1,2,...,n; made one transition to some state k after 'holding' at i for a period τ and then transferred from k to j in the remaining time t - τ, and the probability of this event is $p_{ik} \int_o^t h_{ik}(\tau)\, \phi_{kj}(t-\tau)d\tau$. Summing this latter probability over all possible intermediate states k we obtain

$$\phi_{ij}(t) = \delta_{ij}\tilde{W}_i(t) + \sum_{k=1}^{n} p_{ik} \int_o^t h_{ik}(\tau)\ \phi_{kj}(t-\tau)d\tau \tag{4-10}$$

The convolution under the sign of integration suggests the utilization of Laplace transform theory. Let, as always,

$$f(s) = \int_o^\infty f(t)\ e^{-st}\ dt$$

be the Laplace transform of f(t). Transforming both sides of Eq. (4-10) we obtain

$$\begin{aligned}\Phi_{ij}(s) &= \delta_{ij}\tilde{W}_i(s) + \sum_{k=1}^{n} p_{ik}h_{ik}(s)\ \Phi_{kj}(s) \\ &= \delta_{ij}\tilde{W}_i(s) + \sum_{k=1}^{n} u_{ik}(s)\ \Phi_{kj}(s),\end{aligned} \tag{4-11}$$

where

$$u_{ik}(s) = p_{ik}h_{ik}(s).$$

Equation (4-11) shows two things: <u>first</u>, that the interval transition probabilities $\{\phi_{ij}(t)\}$ are dependent on the <u>product</u> $p_{ij}h_{ij}(s)$, not on the individual quantities. The function $u_{ik}(s)$, derived above in a natural manner is precisely the function $w_{ik}(s)$ defined <u>ab initio</u> in GERT analysis. It plays there the same fundamental role it plays here. <u>Second</u>, that the $\Phi_{ij}(s)$'s are related to each other by <u>a system of linear equations</u>, a typical member of which can be written out as

$$\Phi_{ij}(s) = \delta_{ij}\tilde{W}_i(s) + u_{i1}(s)\ \Phi_{1j}(s) + u_{i2}(s)\ \Phi_{2j}(s) + \ldots + u_{in}(s)\ \Phi_{nj}(s)\ .$$

Recalling what we have previously stated concerning the representation of linear equations by SFGs, this result establishes that SMPs can be represented by SFGs with the generic element as shown in Fig. 4-11. In particular, we must define

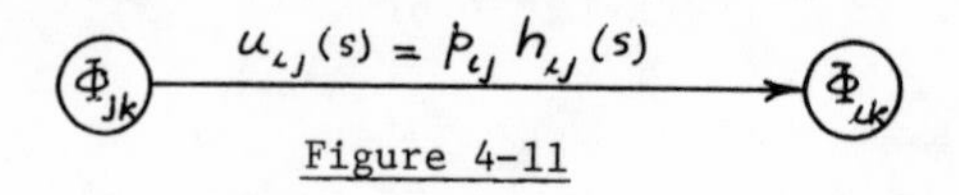

Figure 4-11

n^2 variables $\{\Phi_{ij}(s)\}$, $1 \leq i, j \leq n$, if the probabilities of 'virtual transitions'[†] (p_{ii}) are > 0, or define $n(n-1)$ variables $\{\Phi_{ij}(s)\}$ if $p_{ii} = 0$ for all i. The transmittance $u_{ij}(s) = p_{ij}h_{ij}(s)$ is easily identifiable from the subscripts of the two end nodes of the arc $\Phi_{jk} \rightarrow \Phi_{ik}$, and the direction of the arrows.

The SFG representation of Eq. (4-10) reveals to the naked eye the peculiar structure of this system of linear equations, and simplifies the direct evaluation of the interval transition probabilities by a good measure. To illustrate, consider the simple case of an SMP with only two states, say 1 and 2, on which we define the four interval transition probabilities (dropping the explicit statement of the transform notation (s)): Φ_{11}, Φ_{12}, Φ_{21}, Φ_{22}. The SFG of this system can be drawn with ease, and is shown in Fig. 4-11. Notice that the graph is <u>unconnected</u>: Φ_{11} is related to Φ_{21} while Φ_{22} is related only to Φ_{12}. This is true in general: an n-state SMP can be represented by an SFG composed of n sections, in which the ith section contains only Φ_{ii}, Φ_{ki} for all $k \neq i$, and an 'output node' associated with Φ_{ii}, denoted by 0_i. Each node Φ_{ij}, $i \neq j$, possesses a self-loop of transmittance u_{ii}. The pair of transmittances between Φ_{kj} and Φ_{ij} are determined according to the notation of Fig. 4-10.

The determinant of the complete graph, D of Eq. (4-10), is obviously the product of the sub-determinants of the SFG. In the case of Fig. 4-12, the two sub-determinants are equal, hence

$$D = (1 - u_{11} - u_{22} - u_{21}u_{12} + u_{11}u_{22})^2 = [(1-u_{11})\ (1-u_{22}) - u_{12}u_{21}]^2 .$$

---

† These are the transitions from any state to itself, as distinguished from 'real' transitions which indicate movement to a different state.

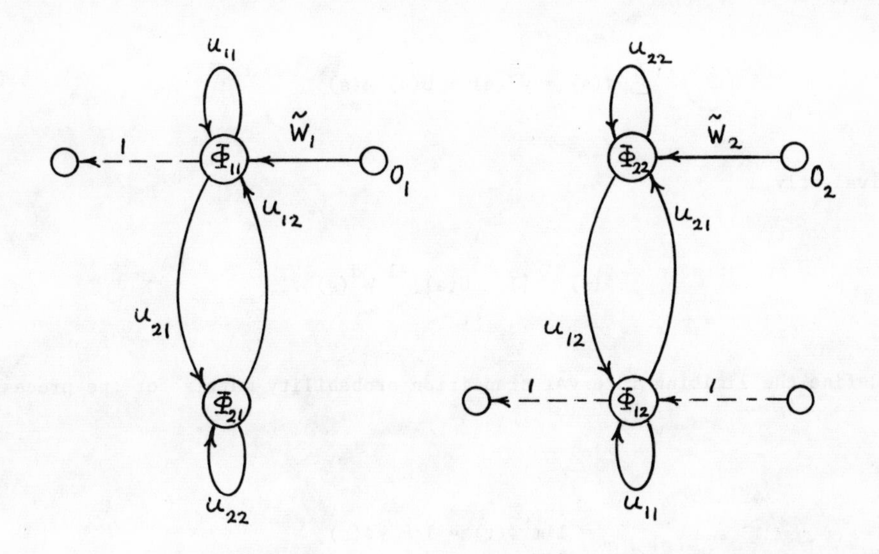

Signal Flow Graph Representation of a 2-State SMP

Figure 4-12

Applying Mason's rule (see Eq. (3-1)) one can immediately write down the expression for each Φ_{ij}. For example

$$\Phi_{11} = \tilde{W}(1-u_{22})/D^{1/2} \quad \text{and} \quad \Phi_{12} = (1-u_{22})/D^{1/2}$$

The transmittance $u_{ij}(s)$ plays a more fundamental role in the analysis of SMPs (or GERT) than that indicated by the above analysis. To show this, it is better to resort to the matrix representation of Eq. (4-10). Let

$\Phi(s)$: the matrix $[\Phi_{ij}(s)]$

$U(s)$: the matrix $[u_{ij}(s)]$

$\tilde{W}^d(s)$: the diagonal matrix whose ith entry is $\tilde{W}_i(s)$

M^d : the diagonal matrix whose ith entry is the unconditional mean holding time $\bar{\tau}_i$ of Eq. (4-7) .

From Eq. (4-9) we have

$$\Phi(s) = \tilde{W}^d(s) + U(s)\ \Phi(s)$$

or, equivalently,

$$\Phi(s) = [I - U(s)]^{-1}\ \tilde{W}^d(s)\ .$$

Let us define the limiting interval transition probability matrix for the process by

$$\Phi_\infty = \lim_{t\to\infty} \Phi(t) = \lim_{s\to 0} s\Phi(s),$$

where the second equality follows from the final value theorem of Laplace transforms. Using Eq. (4-10), we can write this limit in the form

$$\Phi_\infty = \lim_{s\to 0} s\ \Phi(s) = \lim_{s\to 0} s[I - U(s)]^{-1} \lim_{s\to 0} \tilde{W}^d(s)$$

But from the definition of $\tilde{W}_i(t) = 1 - W_i(t) = 1 - \int_0^t w_i(t)dt$, we have that

$$\tilde{W}^d(s) = [I - W^d(s)]/s \qquad (4\text{-}12)$$

where $W^d(s)$ is the diagonal matrix whose ith entry is $w_i(s) = \int_0^\infty w_i(t)e^{-st}dt$.

Hence, $\lim_{s\to 0} W^d(s) = I$, and both numerator and denominator of Eq. (4-12) equal zero. The evaluation of this limit is by L'Hopital's rule,

$$\lim_{s\to 0} \tilde{W}^d(s) = -\frac{d}{ds} W^d(s) \Big|_{s=0}$$

$$= -\frac{d}{ds} \int_0^\infty W^d(t)\, e^{-st}\, dt \Big|_{s=0}$$

$$= \int_0^\infty t\, W^d(t) = M^d$$

where M^d is defined above. Substituting in Eq. (4-11), we get

$$\Phi_\infty = \lim_{s\to 0} s\Phi(s) = \lim_{s\to 0} s[I-U(s)]^{-1} M^d \qquad (4\text{-}13)$$

Therefore, except for the constant diagonal matrix M^d the limiting conditional transition probability is just the limit of the inverse of $s[I - U(s)]$.

It is intuitively clear, as well as can be rigorously proven, that the limiting matrix Φ is of equal rows. Hence, we can talk about ϕ_j, the limiting unconditional transition probability that after a long period of time the system would be in state j. As can be seen from Eq. (4-13), this must be given by the product of a matrix $E = \lim_{s\to 0} s[I - U(s)]^{-1}$, whose rows are identical, and M^d. Howard[9] proved that the jth entry of any row of E is simply

$$e_j = \pi_j / \sum_j \pi_j \bar{\tau}_j \qquad (4\text{-}14)$$

where the vector $\pi = (\pi_1, \pi_2, \ldots, \pi_n)$ is the vector of the steady state probabilities of the imbedded Markov process with transition matrix $\underline{P} = (p_{ij})$. Such is the simple form of these steady-state probabilities.

Needless to say, the majority of GERT are finite state SMPs with one or more absorbing states, and, therefore, interest is usually focused on the probabilities of realization, and the distribution functions, of the random variables associated with the realization of each of the terminal states, such as duration, cost, resource consumption, etc. The approach to these questions is still through the use of the transmittances $u_{ij}(s)$.

Suppose that at time zero the system was in state 1, by definition. For each node $\underline{i} \neq \underline{1}$ let us define the variable $v_i(s)$ as the 'value' of that node in the SFG of transmittances $(U_{ij}(s))$, with $v_1(s) \equiv 1$. The generic element of such a graph is shown in Fig. 4-13. We shall reason recursively as follows.

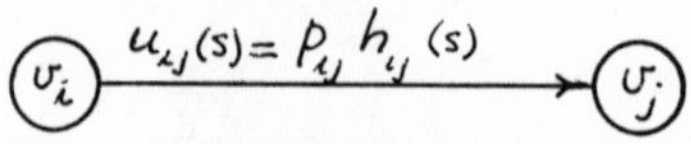

$$(v_i) \xrightarrow{\;u_{ij}(s) = p_{ij}\, h_{ij}(s)\;} (v_j)$$

Figure 4-13

Consider any node connected to node $\underline{1}$, say node $\underline{i}$. By the definition of transmittances we must have

$$v_i(s) = v_1(s) \cdot u_{1i}(s) = u_{1i}(s) = p_{1i}h_{1i}(s).$$

Therefore, the unconditional probability of realization of node $\underline{i}$ is $v_i(0) = p_{1i}$; while

$$m_i(s) = h_{1i}(s) = v_i(s)/p_{1i} = v_i(s)/v_i(0)$$

is the Laplace transform of the p.d.f. of the time to realization of node $\underline{i}$.

On the other hand, if two or more arcs in parallel connect node $\underline{1}$ to node $\underline{i}$, as shown in Fig. 4-14a, then

$$v_i(s) = v_1(s) \sum_k u_{1i}^{(k)}(s) = \sum_k p_{1i}^{(k)}\, h_{1i}^{(k)}(s),$$

from which we deduce that the probability of realization of node $\underline{i}$ is $v_i(0) = \sum_k p_{1i}^{(k)} \leq 1$, and the quantity

$$m_i(s) = \sum_k p_{1i}^{(k)} h_{1i}^{(k)}(s) / \sum_k p_{1i}^{(k)} = v_i(s)/v_i(0) \qquad (4\text{-}15)$$

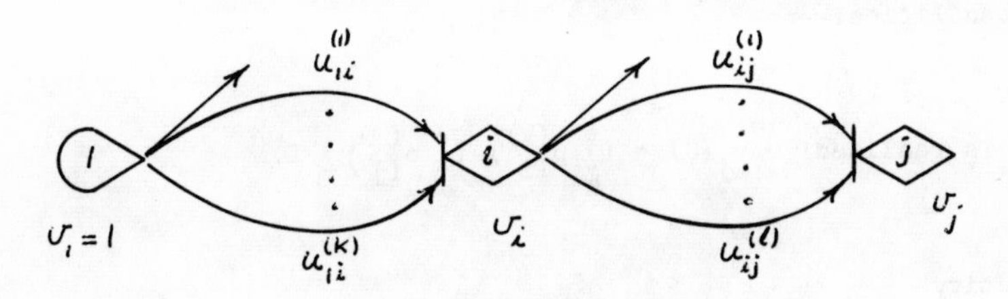

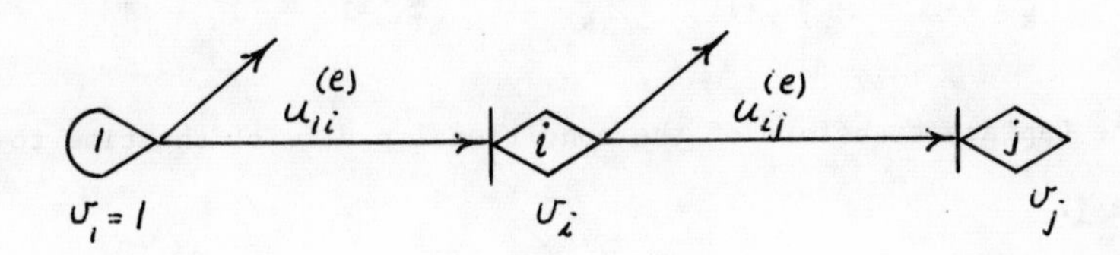

Figure 4-14

is the Laplace transform of the conditional p.d.f. of the time to realization of node i.

Consider next any node j connected to node i by one or more arcs in parallel, as also shown in Fig. 4-14a. Notice that j is two arcs removed from the origin node 1. Again, by straightforward application of SFG rules we have

$$v_j(s) = v_i(s) \sum_{\ell} u_{ij}^{(\ell)}(s)$$

$$= \left(\sum_{k} u_{1i}^{(k)}(s)\right) \left(\sum_{\ell} u_{ij}^{(\ell)}(s)\right).$$

Clearly, the unconditional probability that $\underline{j}$ will be realized is the product of the two probabilities:

$$\Pr(j \text{ is realized}) = v_j(0) = (\sum_k p_{1i}^{(k)}) \; (\sum_\ell p_{ij}^{(\ell)}) \leq 1,$$

and the quantity

$$m_j(s) = (\sum_k u_{1i}^{(k)}(s)) \; (\sum_\ell u_{ij}^{(\ell)}(s))/(\sum_k p_{1i}^{(k)}) \; (\sum_\ell p_{ij}^{(\ell)} = v_j(s)/v_j(0) \qquad (4\text{-}16)$$

is, again, the Laplace transform of the conditional p.d.f. of the time to realization of node $\underline{j}$.

It is now evident that a reduction procedure could have been employed, with equal validity, in which we obtain first the equivalent transmittance u_{1i}^{e} between $\underline{1}$ and $\underline{i}$, and the equivalent transmittance u_{ij}^{e} between $\underline{i}$ and $\underline{j}$ and finally combine the two equivalent transmittances in series into one equivalent transmittance u_{1j}^{e}, as shown in (b) and (c) of Fig. 4-14. The rule can be extended, by iteration, to all the nodes of GERT until the terminal nodes (the absorbing states in the language of SMPs) are reached.

The reader will notice that we have just established that GERT can be analyzed following the rules of SFGs.

In the above discussion we assumed, for simplicity, that node $\underline{1}$ was realized with certainty at time zero, hence $v_1(s) = 1$. Clearly, this assumption can be dropped, and one may generalize the discussion to any node $\underline{i}$ of 'value' $v_i(s)$.

To see the validity of the above statements, consider two arcs (i.e., activities) a and b for which the durations are degenerate: $Y_a = \tau_a$ and $Y_b = \tau_b$ with probability 1. Clearly, we should utilize the z-transform:

$u_a(z) = p_a z^{\tau_a}$ and $u_b(z) = p_b z^{\tau_b}$. Hence, if the two arcs are in series between nodes $\underline{1}$, $\underline{2}$ and $\underline{3}$, with $v_1(z) = 1$, we have

$$v_3(z) = u_a(z) \cdot u_b(z) = p_a p_b z^{\tau_a + \tau_b} .$$

This indicates that the probability of realization of 3 is $p_a p_b$ and that the conditional duration to its realization is a constant $\tau_a + \tau_b$; an intuitively obvious result. On the other hand, if the two arcs were in parallel between nodes 1 and 2, with $v_1(z) = 1$, we have

$$v_2(z) = u_a(z) + u_b(z) = p_a z^{\tau_a} + p_b z^{\tau_b}$$

from which one obtains that the probability of realization of 2 is $v_2(1) = p_a + p_b$ and[†]

$$m_2(z) = v_2(z)/v_2(1) = (p_a z^{\tau_a} + p_b z^{\tau_b})/(p_a + p_b)$$

$$= [p_a/(p_a + p_b)]z^{\tau_a} + [p_b/(p_a + p_b)]z^{\tau_b}$$

which, in turn, implies that the duration of the equivalent arc, Y_e, is a discrete random variable of probability mass function

$$h_e(\tau) = \begin{cases} p_a/(p_a + p_b) & \text{if } Y = \tau_a \\ p_b/(p_a + p_b) & \text{if } Y = \tau_b \\ 0 & \text{otherwise} \end{cases}$$

What if the vector V_a of any arc a contains other parameters beside p_a and Y_a? The answer depends on the behavior of these parameters (which may be random variables): if the parameter behaves in a similar fashion to Y_a then we simply define the Laplace transform (or z-transform) of the parameter in a similar fashion to the way we defined $h_a(s)$. For example, if the cost of an activity is included in V_a, assume it is a random variable γ_a with p.d.f. $g_a(c)$. Clearly,

---

[†] The substitution $z = 1$ in the z-transform is correspondent to the substitution $s = 0$ in the Laplace transform, by reference to the definition of the z-transform.

cost behaves like duration; the costs of two arcs in series add while the cost of two arcs in parallel is the cost of one or the other depending on which activit was realized. Hence, we define

$$g_a(r) = \int_0^\infty g_a(c)e^{-rc}dc$$

and the transmittance of the arc is now

$$u_a(s,r) = p_a h_a(s)\, g_a(r),$$

and analysis proceeds as before. The production shop example given below illustrates this point.

However, if the parameter behaves in a manner different from Y_a, it is not possible to utilize the algebra of SFGs without first transforming the parameter into something which behaves like Y_a. For instance, suppose that the values of the parameter <u>multiply</u> for arcs in series (instead of add as with duration). This is the case of 'explosion' in materials planning, or dis-assembly in production shops, because each unit in activity a generates n_b units in activity b if b follows a, hence, the total number of units for <u>both</u> activities in series is $n_a n_b$. In this case, we first construct the Mellin transform [6] $f(\alpha) = \int_0^\infty x^{\alpha-1} f(x)dx$, then define $u_a(s,\alpha)$ as

$$u_a(s,\alpha) = p_a h_a(s)\, f_a(\alpha)$$

and proceed with analysis as before, since $f(\alpha)$ obeys the same laws of arcs-in-series and in-parallel as the Laplace transforms.

As an example of the application of SFG theory to GANs, consider a shop which produces electric equipment (such as motors, transformers, relays, etc.), and suppose that the flow of a batch of a product can be represented by the GERT of Fig. 4-15a. Basically, the first few steps of manufacture, represented by

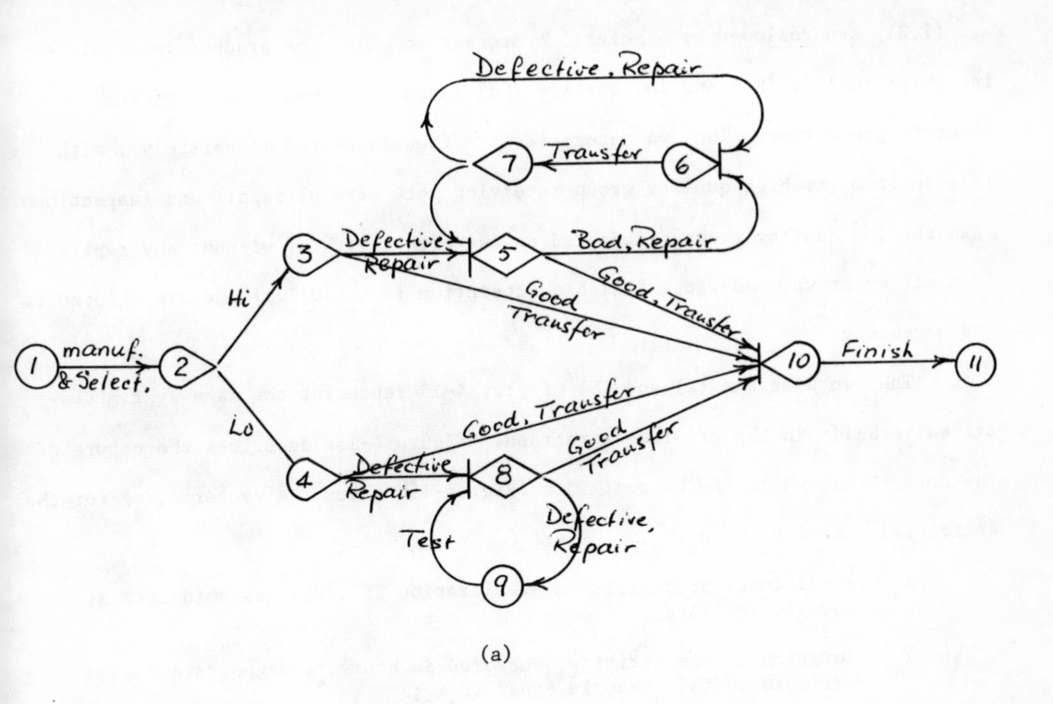

(a)

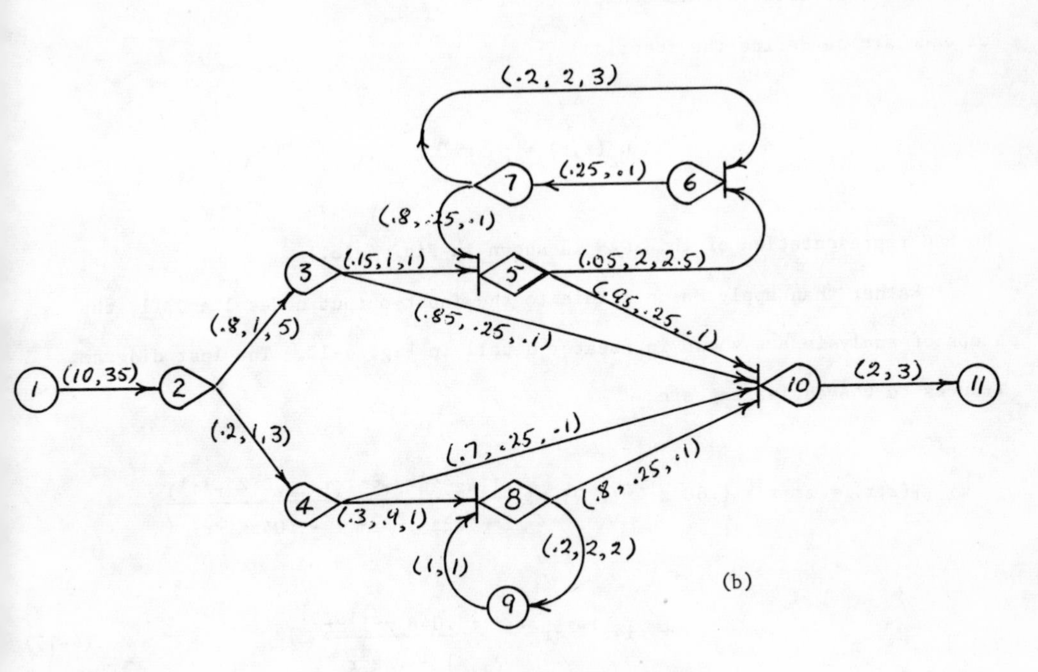

(b)

GERT of an Assembly Operations

Figure 4-15

arc (1,2), are followed by a selection process in which the product is separated into high quality (Hi) and low quality (Lo) through the measurement of some electric properties. The two categories are then inspected (separately), with rejects from the high quality group receiving more careful repair and inspection than the low quality group. Accepted units, either with or without any repair from either group, undergo a finishing operation (arc (10,11)) and are shipped to the warehouse.

The two diagrams (a) and (b) of Fig. 4-15 represent the same system described verbally in the previous paragraph. Figure 4-15a describes the nature of the activities, while the bottom GERT of Fig. 4-15b gives the vectors (V_a) for the activities. Each $V_a = (p_a, Y_a, \gamma_a)$ where

p_a = conditional probability of realization of arc a, assuming its start node is realized;

Y_a = duration of the activity, measured in hours, a degenerate random variable in this example equal to τ_a;

γ_a = cost of the activity, measured in dollars, a degenerate random variable in this example equal to c_a.

For each arc we define the function

$$u_a(z,r) = p_a \, z^{\tau_a} r^{c_a} .$$

The SFG representation of this GAN is shown in Fig. 4-16.

Rather than apply Mason's rule to the input-output nodes 1 and 11, the steps of analysis are shown in greater detail in Fig. 4-17. The last diagram reduces to the equivalent arc

$$u_{1,11}(z,r) = z^{12}r^{38} \{.68\, z^{1.25}r^{5.1} + \frac{.114z^{2.25}\, r^{6.1}(1-.2z^{2.25}r^{3.1})}{1 - .2z^{2.25}r^{3.1} - .04z^{2.5}r^{2.7}}$$

$$+ .14z^{1.25}r^{3.1} + \frac{.048\, z^{2.15}r^5}{1 - .2z^3r^3}\} \qquad (4\text{-}17)$$

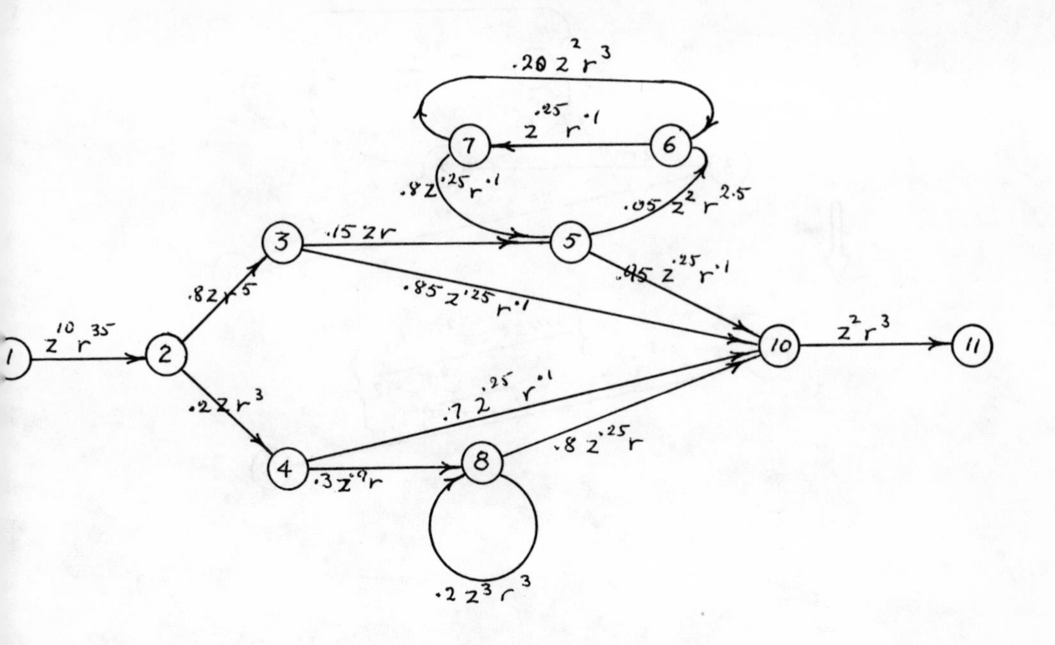

Signal Flow Representation of Fig. 4-15

Figure 4-16

Notice that

$$p_{1,11} = u_{1,11}(1,1) = 1$$

as it should be. Furthermore, one can easily obtain

$$\text{Average processing time} = \frac{d}{dz} u_{1,11}(z,1) \bigg|_{z=1} = 13.51 \text{ periods}$$

$$\text{Average cost of processing} = \frac{d}{dr} u_{1,11}(1,r) \bigg|_{r=1} = \$43$$

Other questions, similar to those previously asked in the analysis of regular

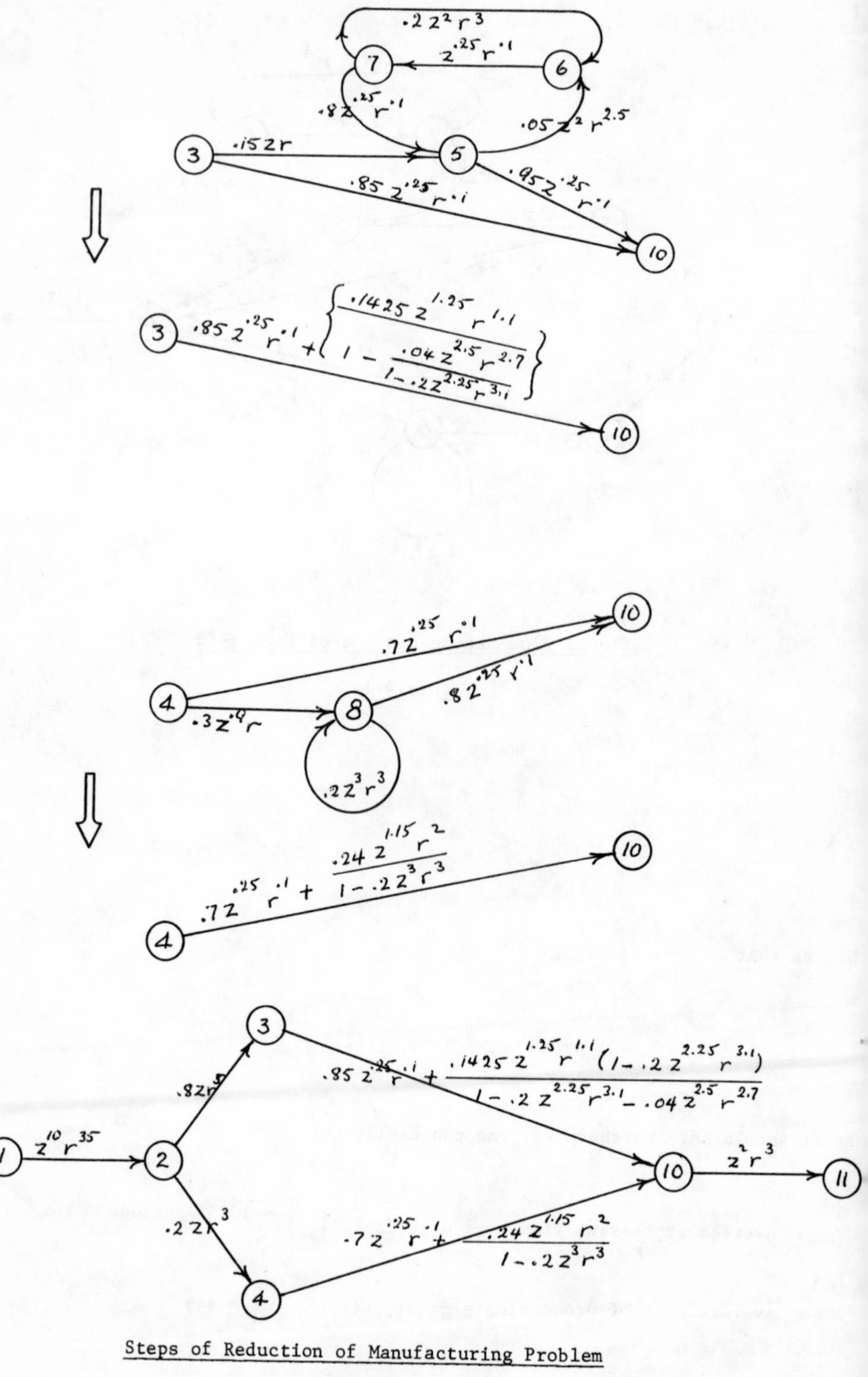

Steps of Reduction of Manufacturing Problem

Figure 4-17

Markov processes, can also be asked and answered in the context of GERT. We shall not pursue these questions here because we would be repeating the same notions but with more complicated expressions, which adds little to insight.

§4.4 ADDENDUM: RE-INTERPRETING THE MACHINE MINDING PROBLEM AS SMP

The reader will recall that in §3 we illustrated the use of SFGs by application to the machine interference problem in §3.2. The illustration involved one repairman and two machines with completely random failure at the rate of λ and completely random repair at the rate μ. Our previous analysis treated the problem directly through representation of the differential-difference equations as an SFG.

In this addendum we re-interpret the queuing problem as an SMP and then represent the latter in SFG style. It is an alternative approach, which should add deeper insight into the problem.[†]

Look at the system in the following light. At any time the system enters a state $\underline{i}$, where i = 0 (all machines working), i = 1 (one machine failed) or i = 2 (both machines down). It immediately chooses the state $\underline{j}$ to which it moves next, with probability p_{ij}. Obviously, from either state $\underline{0}$ or $\underline{2}$ the system can move only to state $\underline{1}$, hence $p_{01} = p_{21} = 1$; while from state $\underline{1}$ it can move to either state $\underline{0}$ or to state $\underline{2}$ with probabilities $p_{10} = \mu/(\lambda+\mu)$ and $p_{12} = \lambda/(\lambda+\mu)$, respectively. Hence, the stochastic matrix of the imbedded Markov process is simply

from \ to	0	1	2
0		1	
1	$\frac{\mu}{\lambda+\mu}$		$\frac{\lambda}{\lambda+\mu}$
2		1	

$$\underline{P} = \text{(matrix above)} \tag{4-18}$$

† For a still different treatment, see ref. [14].

with stationary probabilities

$$\pi_0 = \mu/2(\lambda+\mu) \quad , \pi_1 = 1/2 \; , \pi_2 = \lambda/2(\lambda+\mu) \; . \tag{4-19}$$

Now, the holding time τ_{ij} is a function of the starting state $\underline{i}$ only, and is independent of $\underline{j}$. In particular:

$$h_0(\tau) = 2\lambda e^{-2\lambda\tau}; \; h_1(\tau) = (\lambda+\mu)e^{-(\lambda+\mu)\tau} \; ; \; h_2(\tau) = \mu e^{-\mu\tau}$$

Hence,

$$\begin{aligned} h_0(s) &= 2\lambda/(s+2\lambda) \; ; & \bar{\tau}_0 &= 1/2\lambda \; ; \\ h_1(s) &= (\lambda+\mu)/(s+\lambda+\mu) \quad \text{and} & \bar{\tau}_1 &= 1/(\lambda+\mu) \; ; \\ h_2(s) &= \mu/(s+\mu) & \bar{\tau}_2 &= 1/\mu . \end{aligned} \tag{4-20}$$

The system has three states and, therefore, we need to define nine variables in Eq. (4-10): $\Phi_{00}(s)$, $\Phi_{01}(s),\ldots,\Phi_{22}(s)$. The representation of the relationships among the Φ_{ij}'s as SFG is shown in Fig. 4-18. Notice that since $p_{ii} = 0$ for all states $\underline{i}$, $u_{ii} = 0$ and there are no self-loops in the graph. Furthermore, the zeros in $\underline{P}$ eliminate many arcs in the SFG. We have:

$$\tilde{W}_0(s) = \frac{1}{s}\,[1 - h_0(s)] = 1/(s+2\lambda)$$

$$\tilde{W}_1(s) = \frac{1}{s}\,[1 - h_1(s)] = 1/(s+\lambda+\mu)$$

$$\tilde{W}_2(s) = \frac{1}{s}\,[1 - h_2(s)] = 1/(s+\mu)$$

$$u_{01}(s) = h_0(s) \; ; \; u_{10}(s) = h_1(s)\mu/(\lambda+\mu) \; ; \; u_{12} = h_1(s)\lambda/(\lambda+\mu)$$

$$u_{21}(s) = h_2(s).$$

The remarks leading to Eq. (4-14) lead immediately to the evaluation of the limiting interval transition probabilities, (ϕ_j). Using Eqs. (4-19) and (4-20) we have that

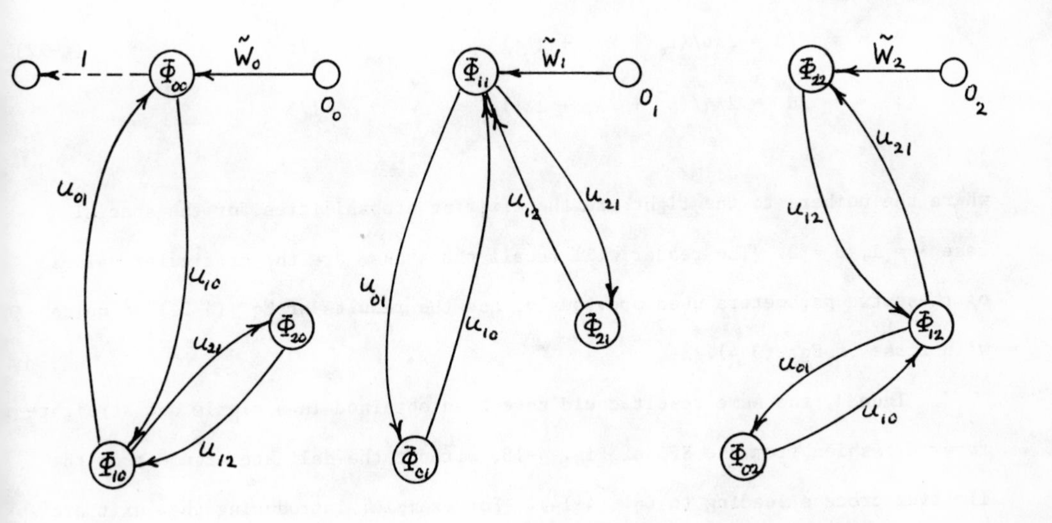

Signal Flow Graph of the Machine Interference

Problem when Interpreted as SMP

Figure 4-18

$$d = \sum_i \pi_i \bar{\tau}_i = (\mu^2 + 2\lambda\mu + 2\lambda^2)/4\lambda\mu\ (\lambda+\mu) \tag{4-21}$$

Hence,

$$\begin{aligned} &\phi_0 = \pi_0\bar{\tau}_0/d = \mu^2/(\mu^2 + 2\lambda\mu + 2\lambda^2) && \Rightarrow && \underline{\lambda = 1,\ \mu = 2} \quad 2/5 \\ &\phi_1 = \pi_1\bar{\tau}_1/d = 2\lambda\mu/(\mu^2 + 2\lambda\mu + 2\lambda^2) && \Rightarrow && 2/5 \\ &\phi_2 = \pi_2\bar{\tau}/d = 2\lambda^2/(\mu^2 + 2\lambda\mu + 2\lambda^2) && \Rightarrow && 1/5 \end{aligned} \tag{4-22}$$

where the numbers to the right are the limiting probabilities for the special case $\lambda = 1$, $\mu = 2$. The reader will recall that these are the particular values of these two parameters used previously, and the results of Eq. (4-22) coincide with those of Eq. (3-4).

Indeed, the same result could have been obtained in a simple and straightforward fashion from the SFG of Fig. 4-18, without the delicate argument of the limiting process leading to Eq. (4-14). For example, introducing the 'exit arc' shown dotted in Fig. 4-18, we immediately have

$$\Phi_{00}(s) = \bar{W}_0 \frac{(1 - u_{21}u_{12})}{(1 - u_{01}u_{10} - u_{12}u_{21})},$$

and

$$\phi_0 = \lim_{s\to 0} s\ \ \Phi_{00}(s) = \lim_{s\to 0} \frac{s[1-\lambda\mu/(s+\mu)(s+\lambda+\mu)]}{(s+2\lambda)[1-2\lambda\mu/(s+2\lambda)(s+\lambda+\mu)-\lambda\mu/(s+\mu)(s+\lambda+\mu)]}$$

Since both numerator and denominator approach zero as $s\to 0$, the limit is evaluated by L'Hopital's rule,

$$\frac{d}{ds}\text{ (numerator)}\Big|_{s=0} = \mu/(\lambda+\mu)$$

$$\frac{d}{ds}\text{ (denominator)}\Big|_{s=0} = (\mu^2 + 2\lambda\mu + 2\lambda^2)/\mu(\lambda+\mu)$$

Consequently,

$$\phi_0 = \mu^2/(\mu^2 + 2\lambda\mu + 2\lambda^2\ ) \Rightarrow 2/5 \quad \text{at } \lambda = 1 \text{ and } \mu = 2,$$

which is the same result obtained before in Eq. (4-22) and Eq. (3-4). Naturally, similar analysis applies to other nodes.

REFERENCES TO CHAPTER 4

[1] Canada, J. R., "Decision Tree Methodology in Capital Project Evaluation", Proc. 18th Annual Conference, A.I.I.E., May 1967.

[2] Clark, C. E., "The Greatest of a Finite Set of Random Numbers", Oper. Res., Vol. 10, No.3, May-June 1962.

[3] Elmaghraby, S. E., "An Algebra for the Analysis of Generalized Activity Networks", Mgt. Sc., Vol. 10, No. 3, April 1964.

[4] ________________,"On the Expected Duration of PERT Type Networks", Mgt. Sc Vol. 13, No. 5, January 1967.

[5] ________________,"On Generalized Activity Networks", Jour. of Ind. Eng., Vol. 17, No. 11, November 1966.

[6] Epstein, B., "Uses of Mellin Transform in Statistics", Ann. of Math. Stat., Vol. 19, pp. 370-379, 1948.

[7] Fulkerson, D. R., "A Network Flow Computation for Project Cost Curve", Mgt. Sc., Vol. 7, No. 2, January 1961.

[8] ______________, "Expected Critical Path Lengths in PERT Networks", Oper. Res., Vol. 10, No. 6, December 1962.

[9] Howard, R., "System Analysis of Semi-Markov Processes", I.E.E.E. Trans. on Military Electronics, April 1964, pp. 114-124. This paper also includes a good list of references to the theory of semi-Markov processes.

[10] Kelley, J.F., Jr., "Critical Path Planning and Scheduling: Mathematical Basis", Oper. Res., Vol. 9, No. 3, May-June 1961.

[11] Malcolm, D.G., J. H. Roseboom, C. E. Clark, and W. Fazar, "Applications of a Technique for Research and Development Program Evaluation", Oper. Res., Vol. 7, No. 5, Sept-Oct. 1959.

[12] Pritsker, A.A.B., and W. W. Happ, "GERT: Graphical Evaluation and Review Technique - Part I. Fundamentals", Jour. of Ind. Eng., Vol. 17, No. 5, May 1966.

[13] ______________, and P. C. Ishmael, "GERT Simulation Program II (GERTS II)", NASA/ERC, NASA-12-2035, Arizona State University, June 1969.

[14] ______________, and G. E. Whitehouse, "GERT: Graphical Evaluation and Review Technique, Part II--Probabilistic and Industrial Engineering Applications", Jour. Ind. Eng., Vol. 17, No. 6, June 1966.

[15] ______________, ________________"GERT: Part III - Further Statistical Results; Counters, Renewal Times, and Correlations", Trans. Ind. Eng., Vol. 1, No. 1, March 1969.

[16] Thomas, Warren, "Four Float Measures for Critical Path Scheduling", Ind.Eng., Oct. 1969.

[17] Whitehouse, G.E., "Extensions, New Developments, and Applications of GERT", Unpublished Ph.D. Thesis, Arizona State University, August 1965.

Lecture Notes in Operations Research and Mathematical Systems

Vol. 1: H. Bühlmann, H. Loeffel, E. Nievergelt, Einführung in die Theorie und Praxis der Entscheidung bei Unsicherheit. 2. Auflage, IV, 125 Seiten 4°. 1969. DM 12,– / US $ 3.30

Vol. 2: U. N. Bhat, A Study of the Queueing Systems M/G/1 and GI/M/1. VIII, 78 pages. 4°. 1968. DM 8,80 / US $ 2.50

Vol. 3: A. Strauss, An Introduction to Optimal Control Theory. VI, 153 pages. 4°. 1968. DM 14,– / US $ 3.90

Vol. 4: Einführung in die Methode Branch and Bound. Herausgegeben von F. Weinberg. VIII, 159 Seiten. 4°. 1968. DM 14,– / US $ 3.90

Vol. 5: L. Hyvärinen, Information Theory for Systems Engineers. VIII, 295 pages. 4°. 1968. DM 15,20 / US $ 4.20

Vol. 6: H. P. Künzi, O. Müller, E. Nievergelt, Einführungskursus in die dynamische Programmierung. IV, 103 Seiten. 4°. 1968. DM 9,– / US $ 2.50

Vol. 7: W. Popp, Einführung in die Theorie der Lagerhaltung. VI, 173 Seiten. 4°. 1968. DM 14,80 / US $ 4.10

Vol. 8: J. Teghem, J. Loris-Teghem, J. P. Lambotte, Modèles d'Attente M/G/1 et GI/M/1 à Arrivées et Services en Groupes. IV, 53 pages. 4°. 1969. DM 6,– / US $ 1.70

Vol. 9: E. Schultze, Einführung in die mathematischen Grundlagen der Informationstheorie. VI, 116 Seiten. 4°. 1969. DM 10,– / US $ 2.80

Vol. 10: D. Hochstädter, Stochastische Lagerhaltungsmodelle. VI, 269 Seiten. 4°. 1969. DM 18,– / US $ 5.00

Vol. 11/12: Mathematical Systems Theory and Economics. Edited by H. W. Kuhn and G. P. Szegö. VIII, IV, 486 pages. 4°. 1969. DM 34,– / US $ 9.40

Vol. 13: Heuristische Planungsmethoden. Herausgegeben von F. Weinberg und C. A. Zehnder. II, 93 Seiten. 4°. 1969. DM 8,– / US $ 2.20

Vol. 14: Computing Methods in Optimization Problems. Edited by A. V. Balakrishnan. V, 191 pages. 4°. 1969. DM 14,– / US $ 3.90

Vol. 15: Economic Models, Estimation and Risk Programming: Essays in Honor of Gerhard Tintner. Edited by K. A. Fox, G. V. L. Narasimham and J. K. Sengupta. VIII, 461 pages. 4°. 1969. DM 24,– / US $ 6.60

Vol. 16: H. P. Künzi und W. Oettli, Nichtlineare Optimierung: Neuere Verfahren, Bibliographie. IV, 180 Seiten. 4°. 1969. DM 12,– / US $ 3.30

Vol. 17: H. Bauer und K. Neumann, Berechnung optimaler Steuerungen, Maximumprinzip und dynamische Optimierung. VIII, 188 Seiten. 4°. 1969. DM 14,– / US $ 3.90

Vol. 18: M. Wolff, Optimale Instandhaltungspolitiken in einfachen Systemen. V, 143 Seiten. 4°. 1970. DM 12,– / US $ 3.30

Vol. 19: L. Hyvärinen, Mathematical Modeling for Industrial Processes. VI, 122 pages. 4°. 1970. DM 10,– / US $ 2.80

Vol. 20: G. Uebe, Optimale Fahrpläne. IX, 161 Seiten. 4°. 1970. DM 12,– / US $ 3.30

Vol. 21: Th. Liebling, Graphentheorie in Planungs- und Tourenproblemen am Beispiel des städtischen Straßendienstes. IX, 118 Seiten. 4°. 1970. DM 12,– / US $ 3.30

Vol. 22: W. Eichhorn, Theorie der homogenen Produktionsfunktion. VIII, 119 Seiten. 4°. 1970. DM 12,– / US $ 3.30

Vol. 23: A. Ghosal, Some Aspects of Queueing and Storage Systems. IV, 93 pages. 4°. 1970. DM 10,– / US $ 2.80

Vol. 24: Feichtinger, Lernprozesse in stochastischen Automaten. V, 66 Seiten. 4°. 1970. DM 6,– / $ 1.70

Vol. 25: R. Henn und O. Opitz, Konsum- und Produktionstheorie I. II, 124 Seiten. 4°. 1970. DM 10,– / $ 2.80

Vol. 26: D. Hochstädter und G. Uebe, Ökonometrische Methoden. XII, 250 Seiten. 4°. 1970. DM 18,– / $ 5.00

Vol. 27: I. H. Mufti, Computational Methods in Optimal Control Problems. IV, 45 pages. 4°. 1970. DM 6,– / $ 1.70

Vol. 28: Theoretical Approaches to Non-Numerical Problem Solving. Edited by R. B. Banerji and M. D. Mesarovic. VI, 466 pages. 4°. 1970. DM 24,– / $ 6.60

Vol. 29: S. E. Elmaghraby, Some Network Models in Management Science. III, 177 pages. 4°. 1970. DM 16,– / $ 4.40

Offsetdruck: Julius Beltz, Weinheim/Bergstr.

Beschaffenheit der Manuskripte

Die Manuskripte werden photomechanisch vervielfältigt; sie müssen daher in sauberer Schreibmaschinenschrift geschrieben sein. Handschriftliche Formeln bitte nur mit schwarzer Tusche eintragen. Notwendige Korrekturen sind bei dem bereits geschriebenen Text entweder durch Überkleben des alten Textes vorzunehmen oder aber müssen die zu korrigierenden Stellen mit weißem Korrekturlack abgedeckt werden. Falls das Manuskript oder Teile desselben neu geschrieben werden müssen, ist der Verlag bereit, dem Autor bei Erscheinen seines Bandes einen angemessenen Betrag zu zahlen. Die Autoren erhalten 75 Freiexemplare.

Zur Erreichung eines möglichst optimalen Reproduktionsergebnisses ist es erwünscht, daß bei der vorgesehenen Verkleinerung der Manuskripte der Text auf einer Seite in der Breite möglichst 18 cm und in der Höhe 26,5 cm nicht überschreitet. Entsprechende Satzspiegelvordrucke werden vom Verlag gern auf Anforderung zur Verfügung gestellt.

Manuskripte, in englischer, deutscher oder französischer Sprache abgefaßt, nimmt Prof. Dr. M. Beckmann, Department of Economics, Brown University, Providence, Rhode Island 02912/USA oder Prof. Dr. H. P. Künzi, Institut für Operations Research und elektronische Datenverarbeitung der Universität Zürich, Sumatrastraße 30, 8006 Zürich entgegen.

Cette série a pour but de donner des informations rapides, de niveau élevé, sur des développements récents en économétrie mathématique et en recherche opérationnelle, aussi bien dans la recherche que dans l'enseignement supérieur. On prévoit de publier

1. des versions préliminaires de travaux originaux et de monographies
2. des cours spéciaux portant sur un domaine nouveau ou sur des aspects nouveaux de domaines classiques
3. des rapports de séminaires
4. des conférences faites à des congrès ou à des colloquiums

En outre il est prévu de publier dans cette série, si la demande le justifie, des rapports de séminaires et des cours multicopiés ailleurs mais déjà épuisés.

Dans l'intérêt d'une diffusion rapide, les contributions auront souvent un caractère provisoire; le cas échéant, les démonstrations ne seront données que dans les grandes lignes. Les travaux présentés pourront également paraître ailleurs. Une réserve suffisante d'exemplaires sera toujours disponible. En permettant aux personnes intéressées d'être informées plus rapidement, les éditeurs Springer espèrent, par cette série de »prépublications«, rendre d'appréciables services aux instituts de mathématiques. Les annonces dans les revues spécialisées, les inscriptions aux catalogues et les copyrights rendront plus facile aux bibliothèques la tâche de réunir une documentation complète.

Présentation des manuscrits

Les manuscrits, étant reproduits par procédé photomécanique, doivent être soigneusement dactylographiés. Il est recommandé d'écrire à l'encre de Chine noire les formules non dactylographiées. Les corrections nécessaires doivent être effectuées soit par collage du nouveau texte sur l'ancien soit en recouvrant les endroits à corriger par du verni correcteur blanc.

S'il s'avère nécessaire d'écrire de nouveau le manuscrit, soit complètement, soit en partie, la maison d'édition se déclare prête à verser à l'auteur, lors de la parution du volume, le montant des frais correspondants. Les auteurs recoivent 75 exemplaires gratuits.

Pour obtenir une reproduction optimale il est désirable que le texte dactylographié sur une page ne dépasse pas 26,5 cm en hauteur et 18 cm en largeur. Sur demande la maison d'édition met à la disposition des auteurs du papier spécialement préparé.

Les manuscrits en anglais, allemand ou francais peuvent être adressés au Prof. Dr. M. Beckmann, Department of Economics, Brown University, Providence, Rhode Island 02912/USA ou au Prof. Dr. H. P. Künzi, Institut für Operations Research und elektronische Datenverarbeitung der Universität Zürich, Sumatrastraße 30, 8006 Zürich.

ISBN 3-540-**04952**-5
ISBN 0-387-**04952**-5